BRINGING STEM
TO THE ELEMENTARY CLASSROOM

W9-DFB-459

BRINGING
STEM

TO THE
ELEMENTARY
CLASSROOM

Edited by
Linda Froschauer

National Science Teachers Association

Arlington, Virginia

National Science Teachers Association

Claire Reinburg, Director
Wendy Rubin, Managing Editor
Rachel Ledbetter, Associate Editor
Amanda O'Brien, Associate Editor
Donna Yudkin, Book Acquisitions Coordinator

ART AND DESIGN
Will Thomas Jr., Director
Joe Butera, Senior Graphic Designer, cover and
     interior design

PRINTING AND PRODUCTION
Catherine Lorrain, Director

NATIONAL SCIENCE TEACHERS ASSOCIATION
David L. Evans, Executive Director
David Beacom, Publisher

1840 Wilson Blvd., Arlington, VA 22201
www.nsta.org/store
For customer service inquiries, please call 800-277-5300.

NSTA is committed to publishing material that promotes the best in inquiry-based science education. However, conditions of actual use may vary, and the safety procedures and practices described in this book are intended to serve only as a guide. Additional precautionary measures may be required. NSTA and the authors do not warrant or represent that the procedures and practices in this book meet any safety code or standard of federal, state, or local regulations. NSTA and the authors disclaim any liability for personal injury or damage to property arising out of or relating to the use of this book, including any of the recommendations, instructions, or materials contained therein.

PERMISSIONS
Book purchasers may photocopy, print, or e-mail up to five copies of an NSTA book chapter for personal use only; this does not include display or promotional use. Elementary, middle, and high school teachers may reproduce forms, sample documents, and single NSTA book chapters needed for classroom or noncommercial, professional-development use only. E-book buyers may download files to multiple personal devices but are prohibited from posting the files to third-party servers or websites, or from passing files to non-buyers. For additional permission to photocopy or use material electronically from this NSTA Press book, please contact the Copyright Clearance Center (CCC) (*www.copyright.com*; 978-750-8400). Please access *www.nsta.org/permissions* for further information about NSTA's rights and permissions policies.

All photos are courtesy of the chapter authors unless otherwise noted.

Library of Congress Cataloging-in-Publication Data
Names: Froschauer, Linda, editor.
Title: Bringing STEM to the elementary classroom / edited by Linda Froschauer.
Other titles: Science and children.
Description: Arlington, VA : National Science Teachers Association, [2016]
Identifiers: LCCN 2016001935 (print) | LCCN 2016010925 (ebook) | ISBN
     9781681400303 (print) | ISBN 9781681400310 (e-book)
Subjects: LCSH: Science--Study and teaching (Elementary)--United States. |
     Technology--Study and teaching (Elementary)--United States. |
     Engineering--Study and teaching (Elementary)--United States. |
     Mathematics--Study and teaching (Elementary)--United States. | Curriculum
     planning--United States.
Classification: LCC LB1585 .B6865 2016 (print) | LCC LB1585 (ebook) | DDC
     372.30973--dc23
LC record available at http://lccn.loc.gov/2016001935

The *Next Generation Science Standards* ("*NGSS*") were developed by twenty six states, in collaboration with the National Research Council, the National Science Teachers Association and the American Association for the Advancement of Science in a process managed by Achieve, Inc. For more information go to *www. nextgenscience.org*.

Contents

Part 2: GRADES PREK–2 (CONT.)

Part 3: GRADES 3–5

Part 3: GRADES 3–5 (CONT.)

Introduction

There have been few opportunities to completely rethink science education in the United States. With the release of *A Framework for K–12 Science Education* (NRC 2012) and the *Next Generation Science Standards* (*NGSS*; NGSS Lead States 2013) we are embarking on massive change that will influence all future generations. One particularly significant result is an increased focus on engineering education. National initiatives call for an increase in both the quality and quantity of engineering content, and the infusion of engineering into the *NGSS* has firmly established engineering as a core component of elementary education. With the addition of engineering to science classrooms, where math and technology have traditionally been included, we now have STEM (science, technology, engineering, and mathematics). STEM marks a significant change; it's much more than an acronym. Teachers are now deeply involved in bringing about the change—but doing so has not been easy.

Evidence shows that 21st-century workers require skills that many graduates do not acquire through formal education. Students need more experiences that provide in-depth knowledge of the STEM disciplines and apply to problem solving. Many view STEM as a way to develop the skills essential for critical thinking, problem solving, creativity, innovation, communication, and collaboration. A study published by the American Society for Engineering Education identified the following characteristics of quality STEM programs (Glancy et al. 2014):

- The context is motivating, engaging, and real world.

- Students integrate and apply meaningful and important mathematics and science content.

- Teaching methods are inquiry based and student centered.

- Students engage in solving engineering challenges using an engineering design process.

- Teamwork and communication are a major focus. Throughout the program, students have the freedom to think critically, creatively, and innovatively, and they have opportunities to fail and try again in safe environments.

Connecting to the *NGSS*

The authors of the chapters, like most teachers, are initiating STEM strategies as they work toward also meeting the requirements of the *NGSS*. Nationally, in some situations, teachers are not involved in the development of strategies and the use of the *NGSS*; in other classrooms, the curriculum is being readjusted over time to meet the needs of the *NGSS*. In still other cases, the *NGSS* is not serving as the standard for science teaching. However, this book provides classroom connections to the *NGSS* through explicit references to the portions of the investigations that meet the *NGSS* performance expectations, science and engineering practices, disciplinary core ideas, and crosscutting concepts. In many cases, seeing how lessons align with specific *NGSS* concepts and strategies will be helpful as you select investigations for your students and develop a continuum of learning, even if you are not focusing on the *NGSS* as your standard.

At the end of most chapters, you will find a table containing some of the featured components of the *NGSS* targeted in that chapter. The table provides some of the connections between the instruction outlined in the chapter and the *NGSS*. Other valid connections are likely; however, space constraints prevent us from listing all possibilities. In the first column, you will see the *NGSS* component. The adjacent column describes the lesson activities that specifically connect to an *NGSS* standard. In some cases, teachers present the lesson to a group of children off grade level with *NGSS* because pre-assessment indicated that students required off-grade knowledge to build the concept. In other cases, schools are readjusting their curriculum as they attempt to align with the *NGSS* and have not shifted the science content.

Note that the materials, lessons, and activities outlined in chapters featuring preK classrooms provide foundational experiences. Science experiences in preK by their nature are foundational and relate to early elements in learning progressions that facilitate later learning in K–12 classrooms. Because the *NGSS* performance expectations are for K–12, the book does not include specific performance expectations for the preK lessons, but it does identify the disciplinary core ideas that are addressed to show the link between those foundational experiences and students' later learning.

Addressing Classroom Needs

An increasing number of publishers, equipment providers, and schools claim they have created strong programs for developing and encouraging STEM. It's critical to find a source of investigations that will provide the important elements needed for an effective STEM experience. STEM should involve problem-solving skills, serve all students equally well, encourage learning across disciplines, promote student inquiry, engage students in real-world problem solving, and expose students to STEM careers. Students should develop skills in communication, problem solving, data analysis, process following, argumentation based on evidence, solution development, and possibly product design. That is a huge list of expectations that cannot be met through a single lesson. Meeting these needs requires a continuum of learning with a focus on elements that will develop the skills and content over time and through many experiences.

The chapters in this book provide lessons that, in combination with additional learning opportunities, can support teachers in developing STEM in their classrooms. They are based on the actual classroom experiences of teachers who provided these learning opportunities for their students. Whether you are just beginning to delve into STEM experiences or you have been building STEM lessons and are now seeking new ideas, *Bringing*

STEM to the Elementary Classroom will provide you with new, interesting, and productive strategies.

The book is organized in grade-level bands. In this way, you can quickly identify strategies that were developed with a specific grade level in mind. As we all know, however, many strategies can be modified to fit the needs of learners at other grade levels. So begin by looking at the topics within grade bands and then expand the search to other grades, while recognizing that the *NGSS* identified for the chapter are for specific grades. The most important decision is based on what is best for your students and the experiences they need to develop conceptual understanding.

You will find a variety of strategies and topics used throughout *Bringing STEM to the Elementary Classroom*, including the following:

- 5-E Learning instructional phases

- Tested, reliable, and even some original design processes

- Pre-assessment strategies and evaluation rubrics

- Data sheets and learning tools that are readily available for immediate printing and use

- Use of technologies—from digital notebooks to three-dimensional printing

- Challenges that relate to real-world problems, such as filtering water, recycling waste, and collecting water

- Design constructions that solve problems, such as a sound proof wall, wind turbines, moving objects, solar ovens, structures to withstand harsh weather, and protection for living things

- Experiences to develop an understanding of technology

You may find the first two chapters of this book a valuable way to start your journey through the many wonderful ideas shared in *Bringing STEM to the Elementary Classroom*. They provide generic strategies that you can use at many grade levels. Chapter 1 is a good introduction to the process of designing lessons, and Chapter 2 will assist you in identifying the misconceptions of students who are new to STEM.

The goal of *Bringing STEM to the Elementary Classroom* is to help children develop an understanding of the many complex components of STEM. I hope you find these pages filled many new strategies that will support your efforts to provide valuable lessons for your students.

Linda Froschauer
Editor, *Bringing STEM to the Elementary Classroom*
NSTA President 2006–2007

References

Glancy, A., T. Moore, S. Guzey, C. Mathis, K. Tank, and E. Siverling. 2014. Examination of integrated STEM curricula as a means toward quality K–12 engineering education. Paper presented at the annual meeting of the American Society for Engineering Education, Indianapolis.

National Research Council (NRC). 2012. *A framework for K–12 science education: Practices, crosscutting concepts, and core ideas*. Washington, DC: National Academies Press.

NGSS Lead States. 2013. *Next Generation Science Standards: For states, by states*. Washington, DC: National Academies Press. *www.nextgenscience.org/next-generation-science-standards*.

PART 1

GRADES PREK–5

How to Develop an Engineering Design Task

Create Your Own Design Activity in Seven Steps

By Chelsey Dankenbring, Brenda M. Capobianco, and David Eichinger

Recently there has been a push for the integration of engineering content and practices into K–12 science education at the state and national levels (NRC 2012; NGSS Lead States 2013). One way engineering education has been included in the elementary science curriculum is through the use of engineering design tasks as a means to teach science content. Although engineering design tasks may benefit students' science learning and increase their interest in both science and engineering, few teachers have received training on how to create their own standards-based engineering design tasks. In this chapter, we provide an overview of engineering and the engineering design process, then we describe the steps we took to develop a fifth grade–level, standards-based engineering design task titled "Getting the Dirt on Decomposition," which will be referred to as "Compost Bin" in this chapter.

Our main goal is to focus more on modeling the discrete steps we took to create and write an original design brief rather than profiling a specific design task. These steps are generalizable and may serve as common procedural steps you can take to successfully create your classroom-based engineering design task that may complement your curriculum and enhance student learning of science through design. To contextualize our writing steps, we use the Compost Bin as our signature, original task and provide helpful guidelines for implementing this task.

What Are Engineering and the Engineering Design Process?

A Framework for K–12 Science Education (NRC 2012) defines *engineering* as "any engagement in a systematic practice of design to achieve solutions to particular human problems" (p. 11). Engineers use their understanding of mathematics and science concepts to find solutions to ill-structured problems. These problems are "messy" because there is not one correct solution or a single path to arrive at a solution. Rather, engineers make a series of trade-offs as they use their knowledge of science and mathematical concepts to arrive at a solution. Engineers devise a solution, either an artifact or a process, to the problem by going through an engineering design process.

Engineering design is an iterative process in which engineers identify a problem, brainstorm ideas, create a design, construct and test a prototype, and redesign their prototype as needed (Capobianco, Nyquist, and Tyrie 2013).

In an elementary science classroom, students can be introduced to an engineering design task through a design brief which describes a client with a real-world problem that needs solving. Students individually develop a plan to solve the problem and document their plan in a design notebook. Next, students are placed into teams of three to four students and share their individual plan with the other members of their team, discussing the strengths and weaknesses of each idea. The design team develops a group design and constructs a prototype using materials readily available in a classroom or home. Each team tests its design and records the data in its design notebook. The results are evaluated based on the criteria and constraints presented in the design brief. Students present their design to the class and explain how the science concepts were used to inform their design. Based on the results of their testing, students redesign their prototype to better meet the client's needs. This process can be modified to include more teacher guidance for early elementary students.

Scientific Inquiry Versus Engineering Design

An important feature of scientific inquiry-based and engineering design-based learning is that the activities are student-centered rather than teacher-centered. In addition, both inquiry and engineering design can be used to effectively teach science concepts and problem-solving practices. Yet key differences exist that make these two approaches very separate entities.

One difference is the starting point. Inquiry activities begin with a question that needs to be answered, whereas design-based activities begin with a problem that needs to be solved. Another difference is the approach taken by the teacher for each type of activity. Approaches to inquiry activities can vary between directed, guided, and open-ended. Each approach depends on the amount of detailed guidance that the teacher provides (NRC 2000). Directed inquiry activities, for example, are often used to validate a scientific idea; therefore, there is a predetermined correct answer for an inquiry activity. Design-based activities, however, are open ended; there are numerous ways to solve the problem. Unlike inquiry where the final product is the answer to the question posed, design ends with a constructed artifact or process (Dankenbring, Rupp, and Capobianco 2013).

Completed compost columns are shown.

Creating Your Own Engineering Design Task

The steps below outline the process we took to create a fifth-grade engineering design task. As we describe each step, we juxtapose the example of our design task, Compost Bin, to illustrate the process we used to develop our own design task. In this task, students are challenged to help the citizens of Haiti create an efficient compost column that yields better compost for fertilizing their crops. By following the steps outlined below, you can develop original design tasks or transform an existing activity into a design task. For a detailed lesson plan of the Compost Bin design task and supporting materials, see Internet Resources (p. 10).

Step 1: Identify a Standard

The first step is to identify a science standard, including the performance expectation and related disciplinary core idea(s). We recommend that each design task focuses on only one or two standards so students can easily identify and incorporate the science concepts being emphasized into their designs. When selecting the appropriate science standard, consideration must be given to whether the standard merits a design-based activity. Performance expectations that require students to "identify" or "describe" a phenomenon are not as friendly to design tasks, whereas more open-ended standards that are conducive to investigating, designing, and examining concepts or phenomena are. Thus, performance expectations that ask students to "investigate," "demonstrate," "design," "observe," or "determine" are likely suitable for a design task. This task aligns with the *Next Generation Science Standards* fifth-grade life science standard, Ecosystems: Interactions, Energy, and Dynamics. This task also addresses the disciplinary core ideas of interdependence in ecosystems and cycles of matter and energy transfer in ecosystems, and it incorporates the crosscutting concept of systems and system models.

Step 2: Brainstorm Ideas for Design-Oriented Activities

The next step is to brainstorm activities that could be adapted into engineering design tasks. A good starting point is listing activities, such as inquiry activities, you or your colleagues use or have used previously to teach the concept. Attention must be given to selecting and focusing on an activity that can become an open-ended problem that students can solve by constructing an artifact or process. Activities that are malleable and provide students the opportunity to test a range of ideas are preferred.

When brainstorming activities related to decomposition, we identified several of our favorite life science activities, including decomposing different fruits, leaves, and trees; observing soil for living and nonliving materials; and constructing compost bins. We selected the Compost Bin because the nature of this activity allows for the development of multiple design tasks. Rather than ask students to design a compost bin, students were provided with a compost bin made from 2 L bottles and were asked to create a process for making compost inside the column (see Internet Resources, p. 10). We encouraged students to conduct research on the different materials required for decomposition to occur, materials that decompose quickly, and the best arrangement of the materials within a compost bin (see NSTA Connection, p. 10).

Step 3: Contextualize a Problem Statement Within the Design Task

This step involves thinking about various everyday problems that relate to the artifact or process that will be constructed during the design activity. The problem must be open-ended and allow students to use a diverse set of ideas and methods to solve the problem through the application of a science concept. When doing this, ask yourself the following questions: What types of everyday problems do we encounter that relate to this science concept? How can we apply our understanding of this phenomenon to benefit society?

1

When developing the problem statement for our compost task, we thought about what decomposition is, why it is important, and what problems it can solve. Decomposition recycles nutrients back into the soil, which plants can then use. Based on this information, we came up with problems that decomposition could possibly help solve: poor soil quality, the amount of materials sent to landfills, and gas emissions associated with landfills. We chose to focus on poor soil quality with the goal of introducing compost as a means of solving the problem.

Step 4: Identify Necessary Materials, Resources, and Tools

Like most activity-based science lessons, consider using materials that are inexpensive and can be found in your classroom or at a local home improvement store. This makes the design tasks affordable and students will be familiar with how these materials can be manipulated and used to construct a prototype. Provide a variety of materials and you will be amazed at what students come up with! Make sure to keep in mind safety concerns when selecting materials for any classroom activity.

We generated a list of materials necessary to construct the column and make the compost. These materials included: 2 L soda bottles, scissors, grass, brown and green leaves, fruit, twigs, red worms, and cheesecloth. Additional materials included cleaned eggshells, plant matter, vegetables, and bread. Students should exercise caution when handling scissors and cutting bottles because the edges may not be smooth. We recommend providing students with a column that has already been constructed with tape covering the cut edges to minimize safety concerns (see the detailed lesson plan in Internet Resources, p. 10, for safety guidelines and a list of materials). We suggest collecting items such as 2 L bottles well in advance of the activity or

asking students to donate materials (e.g., bottles, fruit peels) to class. For this activity, some of the materials are seasonal (e.g., grass, twigs, leaves) and should be collected in the fall, especially if you plan to start the columns during the winter.

Step 5: Develop the Design Brief

Design briefs provide the context of the design task, the background information, and the details necessary for students to devise a solution. You should introduce the problem using an authentic context that includes a goal, client, end user, criteria, and constraints. Take the problem statement you created in step 3 and place it within an authentic context. Ask yourself, "To what context can my students relate?" In other words, is there a problem in our school or community that the students are invested in? Providing an authentic context will keep your students motivated and engaged.

Once you have identified a real-world context, consider a scenario in which an imaginary or authentic client would hire your students to develop a solution to the problem that benefits a group of people—the end users. Next, generate a list of criteria—the desired features of the prototype—that should be met and a list of constraints, or the boundaries students are to work within. Some constraints include time available to design and build, materials available, size requirements for the design, and the cost to construct the prototype.

For our compost task, we chose to focus on a particular country that most students have heard of, Haiti. Haiti is a country that depends on agriculture for food and income, but it has poor soil quality because of drought, soil erosion, and deforestation. The scenario we created for our design brief was Haitian citizens needed help designing a process for making compost that will enhance their soil quality (see NSTA Connection,

p. 10). However, the brief could be easily adapted to a local scenario, such as a local farm, a gardening center, city streets, zoo, or nature center.

Step 6: Implementing the Design Task

Engineering design tasks can require a significant amount of time to implement; students may spend a week or more completing the design task. The compost bin tasks took approximately three to five 40-minute class sessions to research, design, and assemble, then an additional six weeks to monitor (10 minutes per week) and generate actual compost. The first week included the introduction of the design brief, time to research the topic of decomposition, and several 30–40 minute class sessions to assemble the actual column using information students gathered from their research.

Once the columns were assembled, the following five weeks involved students making weekly observations of their columns and gathering data (e.g., soil temperature, appearance, amount of moisture, and odor). As composting results in the growth of bacteria and mold, students with mold allergies should be cautious during data collection and teachers should ensure students wash their hands with soap and water after each session. In the case of severe mold allergies, the compost bins can be stored outside of the classroom and data collection can be done outside. After six weeks, the students share the results and make observations of each team's compost. Students record their reflections on the process and identify one or two ways to improve their approach for making compost. Students may respond to one or more of the following reflection prompts: Was your team successful at creating compost? Describe your observations of your compost (Is it dry, wet or moist? What is the average temperature of the soil?). If you could change the process your teams used to make

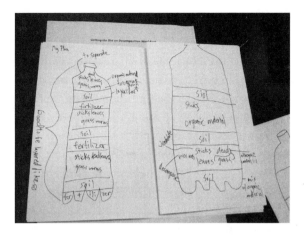

A student's completed notebook entry is shown.

the compost, what kinds of changes would your team make? What kinds of materials (organic or inorganic) would you add?

Step 7: Assessing Students' Engagement in Design

Assessing students' learning is an important feature of an engineering design task. Although our focus in this chapter is on modeling the steps to develop an engineering design–based task, assessment is complementary to the process. When we engage students in an engineering design task, we evaluate their performance on different features of the task. Sometimes we assess how well students work in a design team or maintain a design notebook. Application of science concepts, such as decomposers and biotic and abiotic factors, can be assessed during students' individual design, group design, design construction, and explanation or reflection on the design's performance during testing.

Conclusion

As science teachers begin integrating engineering principles and practices into their curriculum, it is imperative that they feel comfortable creating and implementing engineering design tasks in their classroom. Engineering design tasks provide students with real-world problems situated in authentic contexts. This helps pique students' interest while making connections between science and their everyday lives. As students design, build, and test their prototypes, they are able to challenge their conceptions of scientific phenomena and witness firsthand any flaws in their understanding. Thus, being aware of engineering resources and knowing how to create your own engineering design task are valuable tools for the elementary science teacher.

Acknowledgment

This material is based on work supported by the National Science Foundation under Grant No. 0962840. Any opinions, findings, and conclusions or recommendations expressed in this material are those of the authors and do not necessarily reflect the views of the National Science Foundation.

Connecting to the *Next Generation Science Standards*

The materials, lessons, and activities outlined in this chapter are just one step toward reaching the performance expectations listed below. Additional supporting materials, lessons, and activities will be required.

2-ESS2 Earth Systems www.nextgenscience.org/2ess2-earth-systems **2-PS1 Matter and Its Interactions** www.nextgenscience.org/2ps1-matter-interactions **K-2-ETS1 Engineering Design** www.nextgenscience.org/k-2ets1-engineering-design	**Connections to Classroom Activity**
Performance Expectations	
2-ESS2-1: Compare multiple solutions designed to slow or prevent wind and water from changing the shape of the land	Built models to explain and compare solutions that will slow or prevent wind and rain from changing the soil
2-PS1-1: Plan and conduct an investigation to describe and classify different kinds of materials by their observable properties	Planned and conducted an investigation of the different kinds of earth material that make up the three different soils found near the school
K-2-ETS1-1: Ask questions, make observations, and gather information about a situation people want to change to define a simple problem that can be solved through the development of a new or improved object or tool	Asked questions and made observations about the soil in the urban marsh to determine which design would best prevent the wind and rain from changing the soil
Science and Engineering Practices	
Developing and Using Models	Developed three soil profile models and used the models to compare the soils
Constructing Explanations and Designing Solutions	Used evidence to determine the identity of each "mystery soil"

2-ESS2 Earth Systems *www.nextgenscience.org/2ess2-earth-systems* **2-PS1 Matter and Its Interactions** *www.nextgenscience.org/2ps1-matter-interactions* **K-2-ETS1 Engineering Design** *www.nextgenscience.org/k-2ets1-engineering-design*	**Connections to Classroom Activity**
Science and Engineering Practices	
Planning and Carrying Out Investigations	Planned and conducted an investigation resulting in collection of data about the different materials in soils
Disciplinary Core Ideas	
ESS2.A: Earth Materials and Systems • Wind and water can change the shape of the land.	Investigated and analyzed results and determined how wind and rain can change soils
ESS2.B: Plate Tectonics and Large-Scale System Interactions • Maps show where things are located. One can map the shapes and kinds of land and water in any area.	Used a topographical map and an aerial map to predict where to locate different kinds of soil in the neighborhood
PS1.A: Structure and Properties of Matter • Different kinds of matter exist and many of them can be either solid or liquid, depending on temperature. Matter can be described and classified by its observable properties.	Discovered that soil is composed of different components: organic materials and sand, silt, and clay
ETS1.A: Defining and Delimiting Engineering Problems • A situation that people want to change create can be approached as a problem to be solved through engineering.	Solved a problem based on a design brief identifying resources, constraints, and desired outcomes Researched the problem, tested proposed solutions using a model, and conveyed findings
Crosscutting Concepts	
Patterns	Used patterns seen in soils to consider how soil is the same and different
Stability and Change	Explained how rain and wind can change soil and developed solutions to slow or prevent change
Energy and Matter	Explained how soil looks like one material but actually contains many different materials

Source: NGSS Lead States 2013.

References

Capobianco, B. M., C. Nyquist, and N. Tyrie. 2013. Shedding light on engineering design. *Science and Children* 50 (5): 58–64.

Dankenbring, C. A., M. Rupp, and B. M. Capobianco. 2013. Engineering design in the elementary science classroom. *Hoosier Science Teacher* 39 (1): 25–29.

National Research Council (NRC). 2000. *Inquiry and the National Science Education Standards: A guide for teaching and learning.* Washington, DC: National Academies Press.

National Research Council (NRC). 2012. *A framework for K–12 science education: Practices, crosscutting concepts, and core ideas.* Washington, DC: National Academies Press.

NGSS Lead States. 2013. *Next Generation Science Standards: For states, by states.* Washington, DC: National Academies Press. *www.nextgenscience.org/next-generation-science-standards.*

Internet Resources

Boston Museum of Science's Engineering is Elementary
www.mos.org/eie

Bottle Biology: Build a Decomposition Investigation Column
www.bottlebiology.org/investigations/decomp_build.html

Compost Bin Detailed Lesson Plans
https://stemedhub.org/resources/12

Dayton Regional STEM (Science, Technology, Engineering, and Math) Center
http://daytonregionalstemcenter.org/stem-framework-101

Science Learning Through Engineering Design Partnership
http://stemedhub.org

TeachEngineering
www.teachengineering.org

TryEngineering
www.tryengineering.org

NSTA Connection

For the Design and Build a Compost Column design brief, decomposition research resources, and the rubric, visit **www.nsta.org/SC1410.**

2

Minding Design Missteps

A Watch List of Misconceptions for Beginning Designers

By David Crismond, Laura Gellert, Ryan Cain, and Shequana Wright

Any teacher has a right to feel a bit overwhelmed when asked to include a new suite of learning goals and activities into an already crowded teaching schedule. Add to this a dash of fear when the new subject involves topics you never studied in school or college and a new grab bag of activities and pedagogical techniques.

The *Next Generation Science Standards* (*NGSS*; NGSS Lead States 2013) ask teachers to give engineering design equal standing with scientific inquiry in their science lessons. Students finishing elementary school "should have had numerous experiences in engineering design." What do engineering design practices look like, and how do you assess them? How similar and different is engineering design from scientific inquiry? What sorts of misconceptions are beginning designers prone to, and what can you as their teacher do to help them?

A Framework for K–12 Science Education (*Framework*; NRC 2012) presents practices that are shared by science and engineering (Figure 2.1, p. 12). The *NGSS* describe performance expectations

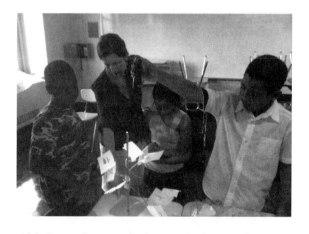

With the goal to cast the largest shadow, students designed a model tree using a wooden dowel, foam block, wire, and sticky notes. They observed real trees as part of their research and then improved their designs.

for engineering design for two elementary grade bands: K–2 and 3–5 (Table 2.1, p. 12).

Looking over a table of sample design activities from well-known curriculum sources might help "make the strange familiar" and show you that you probably have done engineering tasks with your students already (Table 2.2, p. 13).

FIGURE 2.1

The *Framework* Practices

1. Asking questions (for science) and defining problems (for engineering)

2. Developing and using models

3. Planning and carrying out investigations

4. Analyzing and interpreting data

5. Using mathematics and computational thinking

6. Constructing explanations (for science) and designing solutions (for engineering)

7. Engaging in argument from evidence

8. Obtaining, evaluating, and communicating information

Source: NRC 2012.

TABLE 2.1

Comparable Design Performances for K–2 and 3–5 Students in the *NGSS*

Grades K–2	Grades 3–5
K–2-ETS1-1: Ask questions, make observations, and gather information about a situation people want to change to define a simple problem that can be solved through the development of a new or improved object or tool.	**3–5-ETS1-1:** Define a simple design problem reflecting a need or a want that includes specified criteria for success and constraints on materials, time, or cost.
K–2-ETS1-2: Develop a simple sketch, drawing, or physical model to illustrate how the shape of an object helps it function as needed to solve a given problem.	**3–5-ETS1-2:** Generate and compare multiple possible solutions to a problem based on how well each is likely to meet the criteria and constraints of the problem.
K–2-ETS1-3: Analyze data from tests of two objects designed to solve the same problem to compare the strengths and weaknesses of how each performs.	**3–5-ETS1-3:** Plan and carry out fair tests in which variables are controlled and failure points are considered to identify aspects of a model or prototype that can be improved.

Source: NGSS Lead States 2013.

TABLE 2.2

Familiar Engineering Design Activities

Design Task	Curricula and Web Links
A. Create a model parachute that falls slowly.	A Long Way Down: Designing Parachutes (Engineering is Elementary, www.eie.org/eie-curriculum/curriculum-units/long-way-down-designing-parachutes)
	Air and Weather (Full Option Science System, www.fossweb.com/module-summary?dDocName=D995800)
	Diving Into Science (Project-Based Inquiry Science, http://its-about-time.com/pbis/units/dic.html)
B. Design a model car that travels far.	Vehicles in Motion (Project-Based Inquiry Science, http://its-about-time.com/pbis/units/vim.html)
	Fantastic Elastic and EnerJeep (City Technology, http://citytechnology.org/educators-1)
C. Create a model boat that holds the greatest load.	National Science Resources Center. 2002. *Floating and sinking: Teacher's guide.* Burlington, NC: Carolina Biological Supply Company.
D. Design a shopping bag that holds a given load.	Stuff That Works! Packaging (City Technology, http://citytechnology.org/stuff-that-works/publica tion_packaging.html)
	How Could a Carrier Make the Job Easier? (Nuffield Primary Solutions in Design and Technology, www.nationalstemcentre.org.uk/elibrary/resource/818/how-could-a-carrier-make-the-job-easier)
E. Design a model tree that makes the most food from the Sun.	Exploring Trees and Ponds (Educational Development Center, treesandponds.edc.org)

Before presenting a watch list of misconceptions that beginners can fall prey to when designing—useful for you as teacher to know so that you can intervene when you notice students doing them—you, too, might be subject to a few misconceptions of your own regarding engineering design activities.

- *Design tasks rarely have single "right" answers.* There are no right designs for shopping bags, boats, or cars because they are designed for different users and different purposes. A bag for holding recently purchased books must support more weight and be studier than a bag for carrying birthday cards or candles.

- *Design tasks are very different from step-by-step construction activities.* There are kits that your school can purchase that purport to be design activities but which are really step-by-step build-it projects. Good design challenges to varying degrees are more open-ended. They let students know how their final product should perform within given limitations. They do not show, however, what approach should be taken, how a model or prototype should be made, or what it should look like. Such things are for the designers to decide.

Habits of Beginning Designers

Each of the following six habits of beginning designers starts with a reference to the *NGSS* it addresses, gives a description of the habit or misconception, and ends with some suggested teaching strategies that can remedy the misconception.

HABIT #1: Beginning designers get tempted to "build first" before comprehending the challenge.

- *NGSS* science and engineering practice: Asking Questions and Defining Problems

- *NGSS* disciplinary core idea: Defining and Delimiting Engineering Problems (ETS1.A)

Distribute materials for a design challenge to young children, along with a description of the design challenge, called a design brief, and you can bet that some students will start making things without a good idea of what the device should do or how the materials behave. Students need some grasp of the challenge in front of them before they start. This involves the act of reading comprehension, where you make sure that your students know what the product must do before they go to work.

Students need to understand, for example, that a model parachute they are designing should take as long a time as possible to fall a given distance. How this is achieved is up to them. If an indoor catapult that they are designing will be judged by its accuracy, they should know how this can be measured by noting how close to a bulls-eye the projectile hits the target. A useful literacy task would ask students to write a "problem statement" in their own words that tells how their product must perform (criteria) and what limits there are in making that product (constraints).

Note that problem *understanding* is different from problem *framing*. The former involves comprehension while the latter requires a deeper grasp of the problem, such that the designers can describe the approach they want to take. Will the robot by design take the offensive or use defensive play tactics? Will the parachute descend straight down or glide at an angle? Will the rubber-band car by design gain its speed more quickly and over a shorter distance or accelerate more gradually over a longer distance? These are choices designers make well after they have achieved a basic understanding of the task.

Six Habits of Beginning Designers

- Habit #1: They build first before comprehending the design challenge.

- Habit #2: They skip doing research that can help their design plans.

- Habit #3: They get stuck (fixated) on their first design ideas rather than brainstorming to think "outside the box."

- Habit #4: They make design decisions without weighing both the benefits and trade-offs of all ideas being considered.

- Habit #5: They are unfocused when observing prototype tests and need support in troubleshooting their designs effectively.

- Habit #6: They perform design practices once, step-by-step, and always in the same order for all challenges.

HABIT #2: Beginning designers may skip doing the research that can help them with their design thinking.

- *NGSS* science and engineering practice: Obtaining, Evaluating, and Communicating Information

- *NGSS* disciplinary core idea: Developing Possible Solutions (ETS1.B)

Beginning designers mistakenly think that *designing* is the same thing as *inventing*, that everything they design must be from scratch. Experienced or informed designers conduct research to find out how other designers have addressed similar challenges in the past—professional designers call this looking at "prior art."

The Stuff That Works! shopping bag challenge (Benenson and Neujahr 2004) has students do a fun research activity by conducting a scavenger hunt before they design their own shopping bags. Students bring to class their favorite bags of various designs, organize them into piles, and then ask other students to guess the categories they created. The free Nuffield Primary Solutions in Design and Technology curriculum (see Resources) has students do quick investigations of the different ways people carry shopping bags (called carriers in the United Kingdom). They bring into class a collection of bags that they take apart to learn how handles are attached and how bags are reinforced to make them stronger.

Research in the classroom can be as simple as doing a search on the internet or having a student from each team take a quick tour of the classroom to see what other teams are developing. It can also take the form of quick-and-dirty field investigations. Teacher and coauthor Shequana Wright took her kindergarten students on a walking field trip to take photos of the branches and leaves of trees. The patterns that students noticed in nature informed their designing of a model tree that would cast the largest shadow. This activity, adapted from the Education Development Center's Exploring Trees and Ponds program (see Internet Resources, p. 18), addresses the following *NGSS* performance expectations for Kindergarten students:

- K-LS1-1: Use observations to describe patterns of what plants … need to survive.

- K-ESS3-1: Use a model to represent the relationship between the needs of plants … and the places they live.

- K-PS3-2: Use tools and materials to design and build a structure that will reduce the warming effect of sunlight on an area.

HABIT #3: Beginning designers can get stuck on the first design idea they generate rather than brainstorming lots of ideas to think "outside the box."

- *NGSS* science and engineering practice: Designing Solutions

Products that solve problems are the sought-after fruit of engineers and designers, just as explanations of natural phenomena are the pursuit of scientists. Successful product design requires generating and investigating lots of potential ideas—the more the better. Students, however, often stick to the first ideas they propose. You can tell this when their final projects are near carbon copies of their earliest design sketches. This nagging problem, which afflicts even expert engineers, is called "idea fixation."

Brainstorming is a trademark practice of engineers and involves them in proposing as many different ideas as possible (quantity over quality) while not criticizing ideas as they get put forward. Not judging ideas made when brainstorming is critical for many younger students—they need to feel that it is safe to explore being creative and proposing the unexpected.

The grades 3–5 performance expectations for engineering design ask students to "generate and compare multiple solutions to a problem" (3–5-ETS1-2; NGSS Lead States 2013). Brainstorming can help designers, young and old, get out of the rut of seeing a problem and its potential solutions in the same old way. If you help students generate lots of ideas, they may find it harder (although not impossible) to fixate on any single plan.

HABIT #4: Beginning designers make design decisions without weighing both the benefits and trade-offs of ideas being considered.

- *NGSS* science and engineering practice: Engaging in Argument From Evidence; Analyzing and Interpreting Data

Beginning designers often make design choices without weighing the good and bad of each alternative on the table. They tend to focus on the positives in the plans they like and the drawbacks of plans they do not favor. As a teacher, you can get older students into the habit of articulating both the good and bad sides of each idea under consideration using evidence they collect from prototype tests that they run.

The last engineering design performance expectation (see Table 2.1, p. 12) asks K–2 students to compare the strengths and weaknesses of two devices made to solve the same problem. You can help students meet this expectation by having them do *Consumer Reports*–style product comparisons using familiar items or products. One such classic investigation involves comparing different brands of paper towels to determine their overall quality.

Grade 3–5 students, however, can be asked to go further and plan fair-test experiments that compare different brands of the same product, or different prototypes that have been made to solve the same challenge. In both cases, students need to answer the question: What do you measure, and how, in determining the overall quality of the products? Students should also discuss the conditions they need to control as they test various product brands or compare prototypes that different team members have proposed or developed over multiple design iterations.

HABIT #5: Beginning designers can be unfocused when observing tests of their prototypes and need help in troubleshooting their designs effectively.

- *NGSS* science and engineering practice: Analyzing and Interpreting Data

Students testing their prototypes can often miss flaws in their performance that you will immediately see. Whether they are observing model parachutes fall (see the Design in the Classroom website in Internet Resources, p. 18) or model windmills spin when a fan is placed in front of them, students can either notice or miss key events during prototype tests. They will think they have observed a wonderful test of their prototype, whereas actually they did not.

Help students do effective troubleshooting by having them use a three-step question sequence:

1. What problem or issue have you noticed? (observe and diagnosis)

2. Why did it happen? (explanation)

3. How would you fix the problem? (remedy)

Students may need instruction in doing design-based troubleshooting, especially when they need to zoom in their attention to identify problems (Crismond 2013; Crismond and Adams 2012). Getting students to make and replay videos of prototype tests can help them "see more" from such testing and improve their troubleshooting thinking.

HABIT #6: Beginning designers think design practices should be performed only once, step-by-step, and always in the same order.

- *NGSS* science and engineering practice: Constructing Explanations and Designing Solutions

Beginning designers are inclined to design on "impulse," where they jump from one task and strategy to another with little awareness of what they are doing or why. *NGSS* addresses this tendency by presenting the same 3-step design model across all grades:

- Defining the problem
- Developing possible solutions
- Improving designs

When beginners encounter these models, however, they sometimes think that the design practices mentioned should be done once, in an unchanging order, step-by-step. These learners miss two essential points: Designers do these steps in any order, as needed, and they learn the most by building and testing prototypes to discover flaws and then iteratively improve them.

Designers, even the most expert, rarely design effective solutions on the first try. They learn mainly through doing revisions to original plans and going through the process a number of times

to improve the product. Have students keep their old prototypes so that they can compare them with newer ones and can tell others what they changed and why. For time-strapped teachers, perhaps the greatest learning opportunity that most often gets lost happens when students are given only enough time to make their first prototypes. Most learning with design-based science challenges comes when students iterate and complete the design cycle a number of times.

Conclusion

At times, you may feel overwhelmed when using design tasks with children and also when teaching them about design practices. When this happens, consider that in your day-to-day work as a teacher, you yourself are acting as a designer when you do the following:

- Define the goals of a lesson and the needs of your users (students).

- Adapt and redesign lessons from existing curriculum.

- Create your own discussion questions.

- Devise formative assessment items, quizzes, and tests.

- Troubleshoot lessons that do not seem to be working.

Whether you are encouraging your students to use design practices or trying to connect to your own work as teacher-as-designer, having a watch list of beginning designer habits can help you in speeding up students' growth in design capability. While design tasks can be used as engaging filler activities, when done well, they can show students how science ideas can be helpful in addressing problems that they face in their daily lives.

References

Benenson, G., and J. Neujahr. 2004. *Packaging and other structures.* Portsmith, NH: Heinemann.

Crismond, D. 2013. Troubleshooting: A bridge that connects engineering design and scientific inquiry practices. *Science Scope* 36 (6): 74–79.

Crismond, D., and R. Adams. 2012. The informed design teaching and learning matrix. *Journal of Engineering Education* 101 (4): 738–797.

National Research Council (NRC). 2012. *A framework for K–12 science education: Practices, crosscutting concepts, and core ideas.* Washington, DC: National Academies Press.

NGSS Lead States. 2013. *Next Generation Science Standards: For states, by states.* Washington, DC: National Academies Press. *www.nextgenscience.org/next-generation-science-standards.*

Internet Resources

Design in the Classroom
www.designintheclassroom.com
This website presents classroom videos of design tasks.

Education Development Center: Exploring Trees and Ponds
http://treesandponds.edc.org

Nuffield Design and Technology Curriculum
www.nationalstemcentre.org.uk/elibrary/resource/818/how-could-a-carrier-make-the-job-easier
This website provides a free U.K. curriculum.

World in Motion
www.awim.org/curriculum

3

Teaching Through Trade Books

Design Dilemmas

By Christine Anne Royce

Through two different stories, students are introduced to the process—including the frustrations—of designing something to solve a problem. The experiences of the books' characters are brought into the classroom by having students engage in an engineering and design process. The activities support the development of these practices and assist students in acquiring and applying scientific knowledge.

Trade Books

The Most Magnificent Thing
By Ashley Spires
ISBN:
978-1-55453-704-4
Kids Can Press, 2014
32 pages
Grades preK–2

Synopsis

The young girl in the story demonstrates that making the perfect thing can be challenging. Although the reader does not know what the perfect thing is until the end of the book, the girl demonstrates the trial, error, perseverance, and adaptation aspects of the design process.

Papa's Mechanical Fish
By Candace Fleming
Illustrated by Boris Kulikov
ISBN:
978-0-374-39908-5
Farrar, Straus, and Giroux, 2013
40 pages
Grades 2–4

Synopsis

This is a fictional story based on true events and highlights the life of inventor Lodner Phillips. Papa (the story is told from the perspective of his daughter) tinkers and tries to develop a variety of things but is rarely successful. Finally, he designs a mechanical fish—a submarine that takes friends and family beneath Lake Michigan.

Curricular Connections

Younger students are introduced "to 'problems' as situations that people want to change" (NGSS Lead States 2013). They "[a]sk questions, make observations, and gather information about a situation people want to change to define a simple problem that can be solved through the development of a new or improved object or tool" (performance expectation ETS1-1: Engineering Design; NGSS Lead States 2013). However, "before beginning to design a solution, [students do need to clearly] understand the problem" (disciplinary core idea ETS1-A: Defining and Delimiting Engineering Problems; NGSS Lead States 2013) and are not expected to come up with original solutions necessarily. The "[e]mphasis is on thinking through the needs of goals that need to be met, and which solutions best meet those needs and goals" (NGSS Lead States 2013). One key thing to note is that the engineering and design process is not simply allowing young students to freely build; they need to be working to solve a definite problem.

At the upper elementary grades, engineering design "engages students in more formalized problem solving … that define[s] a problem using criteria for success and constraints or limits of possible solutions" (NGSS Lead States 2013). In the design challenge, students are realizing that "[p]ossible solutions to a problem are limited by available materials and resources (constraints). The success of a designed solution is determined by considering the desired features of a solution (criteria)" (disciplinary core idea ETS1-A; NGSS Lead States 2013).

Because this process is most likely new to students, keep in mind that improvement of and reflection on solutions (even if the solution does not yield a positive outcome) does meet the standard and helps students in their learning progressions.

Grades K–2: Magnificent Things Solve Problems

Purpose
Students will solve a problem through the design and engineering process.

Engage
Share with students either a picture of a snow shovel and a digging shovel or actual shovels, and identify each for them. Ask them to brainstorm why these two shovels look different and when they would use each. Ask them to continue to brainstorm why the shape of each shovel helps to solve the problem of either digging a hole or removing snow. A question to have the students consider at this point is "Do different things with different designs help to solve different problems?" After students have had an opportunity to discuss this idea, open *The Most Magnificent Thing* to page 30 and show only that page with the statement bubbles to the students. On this page, people are looking at a variety of different things that the young girl has made and identifying what each can do. Ask students what they think is happening in this picture, and allow them to elaborate on what they see. Follow up their answers with questions that help them understand that different things help solve different problems people have in everyday life (*Common Core State Standards, English Language Arts* (*CCSS ELA*): Reading Standards for Informational Texts K–5, Integration of Knowledge and Ideas; NGAC and CCSSO 2010). Return to the beginning of this book and read the entire book to the students, stopping at points to ask the following questions:

- What are some of the steps the girl goes through in designing her magnificent thing? (knowing what she wants, sketching it, making it, changing it, using it)

- Why do you think she almost gave up in making the most magnificent thing? What did she do to calm down? (She was mad that it wasn't working out, she stepped away, and she took a walk.)

- On page 30, why were the things exactly what the people wanted in the picture? (The things helped to solve a problem they had.)

- What problem do you think the girl is trying to solve? (She's trying to design a sidecar for her dog to attach to her scooter. This is shown on the last page of the book.)

Materials
The following materials are used for this activity:

- *The Most Magnificent Thing*

- Pictures of a digging shovel and a snow shovel or actual shovels

- Student data sheet (see NSTA Connection, p. 29)

- Matchbox car

- Toy pickup or dump truck

- Marbles or tennis balls

- Pieces of wood to serve as a ramp

- Various types of paper (different weights and lengths) to construct the bridge

- 2 in. × 4 in. pieces of wood to serve as the foundation pillars of the bridge

- Straws or craft sticks

- Masking tape

- Foil

- Plastic wrap

- Other materials that could be used to build an object to contain the marbles or tennis balls

- Safety glasses

Explore
Return to the idea of the shovels, and discuss the idea that sometimes people have a problem they need to solve and may need to design or create something to solve that problem. Present one of the following problems to the students:

- Move a car from one point to another, with the points being spread approximately 20 cm apart. The solution will result in different types of bridges being constructed by students.

- Keep tennis balls or marbles from falling out of a toy vehicle that must go up a ramp and then down the ramp.

At this point, you should make a choice between which problem task will be presented to all students during the Explore stage. They will do the second problem during the Elaborate stage. Ask students to work in groups of three or four students to develop a solution to the problem you presented. Younger students may need classroom aides to help facilitate the process through questions or suggestions. Once a problem has been identified, ask students to discuss where in the real world they may see a similar problem and how it has been solved there. Have them examine the different materials they have to solve the problem and sketch on their student data sheet what their solution will look like (*CCSS ELA*: Writing Standards K–5, Text Types and Purposes; NGAC and CCSSO 2010). Have them build their design and then test it, note if it worked or failed, and add any other observations they made. Ask the students to return to their groups and discuss how they could improve their design. Students will repeat the process of sketching or explaining the design changes on the student data sheet and testing their new design. Depending on the age level, students may want to revise once or twice after the initial trial.

Explain

After the trial and revision process, ask each group to reconstruct their best solution and identify why they selected it. Have each group present their best solution to the class and demonstrate their solutions to the problem as well as explain to the class what makes this solution their best choice (*CCSS ELA*: Speaking and Listening Standards K–5 – Presentation of Knowledge and Ideas; NGAC and CCSSO 2010). After all groups have shared their solution, bring the class together to discuss the different best solutions. How are they similar? Different? Can part of one idea be combined with part of another idea? Why would this improve the design? Return to the story and ask the students to explain how the young girl changed her design over several of the trials (see last page of the book). Compare their process of solving a problem to her process.

Elaborate

Now that the students have had an opportunity to try an engineering problem and go through the process, give them a second problem. Ask them to repeat their experiences of discussion, questioning, observations, sketching a potential solution, testing that solution, and adjusting the design with the second problem. Younger students may need more adult guidance than older students. Being creative and fluid in their thinking will take practice, and providing different opportunities will help students develop thinking skills in this area.

Evaluate

In addition to being able to see the progression of the designs to solve the problem from the illustrations on the student data sheets, you can follow up on what the students were thinking and feeling as they went through the process, as some students may feel the frustration similar to that of the young girl in the story. It is important that students discuss why they were feeling this way and be reassured that sometimes finding the solution to a problem takes time. It is important that they feel comfortable returning to similar situations in the future. Ask the students to either write or illustrate one way they are similar to the young girl in the story on the back of their student data sheet.

Grades 3–5: STEM's Own Amazing Race

Purpose

Students will participate in a design challenge that requires them to construct a boat that can cross a distance of water the quickest.

Engage

Show the cover art of *Papa's Mechanical Fish* to the class and ask them to make predictions about what the story might be about. Once students have developed the idea of an underwater boat or submarine, ask them to consider why someone might want to travel in a submarine. Share the book with the students, making sure to stop at the individual two-page spreads and discuss not only the words but also the information conveyed in the illustrations (*CCSS ELA*: Reading Standards for Informational Texts K–5, Integration of Knowledge and Ideas; NGAC and CCSSO 2010). Questions that can be asked throughout the book to help generate connections to the story include the following:

- Have you ever invented something that worked or that didn't work, like Papa did in the story?

- Why does Papa want to build a mechanical fish, or, in other words, what problem is he trying to solve? (He has wondered what it would be like to be a fish.)

- Why did Papa change his design several times? Using the sketches provided in the story, discuss some of the changes that were made. Explain why each change or modification was made.

For the last question, create on the board or on chart paper a T-chart that identifies the change made and the reason the change was needed. Helping students consider the answer to these questions also starts to set up an understanding of the need to approach engineering design in a more formalized process. It should be noted that similar to the problem presented in this story, not all problems are the same for each person. Different students may see different problems for different situations.

Materials

The following materials are used for this activity:

- *Papa's Mechanical Fish*
- Chart paper
- Markers
- Student data sheet (see NSTA Connection, p. 29)
- Safety goggles
- Stopwatch
- Straws
- 4 marbles per group
- Large plastic "under the bed" bin
- Pictures of different sail boats
- Foil
- White paper
- Newsprint paper
- Tissue paper
- Craft sticks
- Skewers
- Paper cups
- Other building materials that are available

Explore

Explain to students that they will be putting on Papa's hat, so to say, but the problem they are presented with is building a boat that travels across a distance of water the fastest with the criteria and constraints presented in the problem card (see "STEM's Own Amazing Race," p. 24).

The method by which the boat will be propelled will be wind, which will be created by one team member blowing through a straw while remaining at the starting point (departing shore of the channel). Materials to build the boat can truly be anything, but paper construction is the easiest. After looking at pictures of boats, particularly sailboats, have students begin to design their boats.

Ask the students (in teams of four) to first think about how they would design the boat and then to sketch it out on their student data sheet (*CCSS ELA*: Writing Standards K–5, Text Types and Purposes; NGAC and CCSSO 2010). After each student has had a chance to consider what they personally would do, ask them to share their designs with the team and collaborate on which parts of the boat design they want to include in the overall team design and why. Before actually building the boat, students should have discussed several possible design solutions and the positives and negatives they see with each design feature. They should agree on one design that is a combination of the best parts from each individual design and then sketch that design on their student data sheets and label what each part of the boat is intended to do. As students are designing their boats, circulate the room asking them why they are selecting a certain design or material, what they think a particular part of the design will do, and any other questions that relate to the criteria of the problem design or the constraints of the problem (*CCSS ELA*: Vocabulary Acquisition and Use; NGAC and CCSSO 2010). Allow the students to design their boats.

Students should then be allowed to test their initial design without penalty. Whereas an unsuccessful attempt would penalize the team in a real race, the design process involves generating and testing solutions so that students learn to optimize the design. An ideal situation would be to allow one initial design and two revisions to the design. Each revision would require students to note on their student data sheet what they changed on their design and why. Encourage students to change only one aspect at a time to determine what results each change produces.

Explain

Let the race begin! After students have tested and revised their designs, let each team present their boat design to the class and explain the following points (*CCSS ELA*: Speaking and Listening Standards K–5, Presentation of Knowledge and Ideas; NGAC and CCSSO 2010):

- The features on the boat and what benefit they bring to the design
- What changes they made to their initial design and why (can use their sketches to help explain the changes as well as any prototypes)
- Any challenges they still have with the design

STEM's Own Amazing Race

Problem: You and your teammates are participating in a treasure hunt and are stranded on one side of a channel but must make it to the other side to continue the treasure hunt. You are competing against other teams in a way that is similar to the *Amazing Race* television show. There are four people, including yourself, on your team, who will be represented by marbles. You have at your disposal the building materials provided to construct a boat that will take you and your teammates all at once across the channel.

Criteria: Design a boat that will carry you and your teammates (a total of four marbles) across the channel in the fastest amount of time. The boat must float. Only the materials provided can be used to construct the boat.

Constraints: Boats can be propelled only with the wind that one person generates by blowing through a straw and remaining on the shore of the channel from which you depart. Boats must be no larger than 6 in. (15 cm) × 6 in. (15 cm). There is no constraint for height.

Once each group has presented their design to the class, allow them to engage in the race by identifying which student will be responsible for creating the wind by blowing through a straw. Have that student stand at the narrow edge of a shallow plastic storage container (40 in. × 20. in. × 6 in.) that has been filled two-thirds with water. When the boat is placed in the water, they should bend over and blow through the straw only to create wind to move the boat forward. They are not allowed to move to a different edge of the bin. You, the teacher, should serve as the official time keeper and starter for each team trial. Times should be recorded on a piece of chart paper. Once a boat sinks, tips over and cannot go any further, or reaches the other side, the time should stop and the attempt should be labeled as successful or unsuccessful. The fastest time for a successful run wins the race and makes it to the treasure. Each group should be making notes about what design features they thought worked best from other teams. Have students identify what design features created positive aspects in the design and how they would incorporate those features into their design on a different trial.

Elaborate

Now that the students have had a chance to practice the design and revision process, have them become evaluators of someone else's design by asking them to watch a video on middle school cardboard boat races (see Internet Resources, p. 29). Play the video, allowing the students to see all of the different boat races, and then go back and play it again. Be sure to pause the video to discuss each boat. Ask students what worked in that design and what they would change if they were the boat designers. Why do they think a design was successful or unsuccessful?

Evaluate

Through their designs and sketches, explanations throughout the process, and presentations to the class, you should be able to determine if students understand the ideas of design, criteria, constraints, testing, and revision. Those ideas are the main focus of this lesson and not necessarily accomplishing the problem task presented (*CCSS ELA*: Vocabulary Acquisition and Use; NGAC and CCSSO 2010). A final evaluation measure would be to ask the students to turn their student

data sheet over and describe in a few sentences the similarities and differences between the experience Papa had in the story *Papa's Mechanical Fish* and their own experience in designing a boat for the challenge (*CCSS ELA*: Reading Standards for Informational Texts K–5 – Key Ideas and Details; NGAC and CCSSO 2010).

Connecting to the *Common Core State Standards*

This section provides the *Common Core State Standards* (*CCSS*; NGAC and CCSSO 2010), *English Language Arts* (*ELA*), addressed in this chapter to allow for cross-curricular planning and integration. The CCSS state that students should be able to do the following at grade level.

Reading Standards for Informational Texts K–5: Integration of Knowledge and Ideas

- Grade 1: Use the illustrations and details in a text to describe its key ideas.

- Grade 3: Use information gained from illustrations (e.g., maps, photographs) and the words in a text to demonstrate understanding of the text (e.g., where, when, why, and how key events occur).

Reading Standards for Informational Texts K–5: Key Ideas and Details

- Grade 4: Refer to details and examples in a text when explaining what the text says explicitly and when drawing inferences from the text.

Writing Standards K–5: Text Types and Purposes

- Grade K: Use a combination of drawing, dictating, and writing to compose informative/explanatory texts in which they name what they are writing about and supply some information about the topic.

- Grade 2: Write informative/explanatory texts in which they introduce a topic, use facts and definitions to develop points, and provide a concluding statement or section.

- Grade 4: Write informative/explanatory texts to examine a topic and convey ideas and information clearly.

Vocabulary Acquisition and Use is one of the standards for language. This particular standard is across grade levels and states, "Determine or clarify the meaning of unknown and multiple-meaning words and phrases based on grade [appropriate] reading and content."

Speaking and Listening Standards K–5: Presentation of Knowledge and Ideas

- Kindergarten: Add drawings or other visual displays to descriptions as desired to provide additional details.

- Grade 1: Add drawings or other visual displays to descriptions when appropriate to clarify ideas, thoughts, and feelings.

Furthermore, the *CCSS ELA* provide a standard related to the Range of Text Types for K–5. It indicates that students in K–5 should apply the reading standards to a wide range of texts, including informational science books.

Connecting to the *Next Generation Science Standards*

The materials, lessons, and activities outlined in this chapter are just one step toward reaching the performance expectations listed below. Additional supporting materials, lessons, and activities will be required.

K–2-ETS Engineering Design www.nextgenscience.org/k-2ets-engineering-design	Connections to Classroom Activity
Performance Expectations	
K-2-ETS1-1: Ask questions, make observations, and gather information about a situation people want to change to define a simple problem that can be solved through the development of a new or improved object or tool	Identified, asked questions about, and acknowledged a problem
K-2-ETS1-2: Develop a simple sketch, drawing, or physical model to illustrate how the shape of an object helps it function as needed to solve a given problem	Sketched solution to the problem and revised the sketch after testing the solution throughout the design process
Science and Engineering Practices	
Asking Questions and Defining Problems	Asked questions of each other and the teacher while working to solve a problem
Developing and Using Models	Created two-dimensional sketches to serve as a model for their solution

K–2-ETS Engineering Design www.nextgenscience.org/k-2ets-engineering-design	Connections to Classroom Activity
Disciplinary Core Ideas	
ETS1.A: Defining and Delimiting Engineering Problems • A situation that people want to change or create can be approached as a problem to be solved through engineering. • Asking questions, making observations, and gathering information are helpful in thinking about problems. • Before beginning to design a solution, it is important to clearly understand the problem.	Presented with two different problems throughout the entire lesson that need to be solved Asked questions of each other and the teacher while working to solve a problem Verbalized the problem and what a successful solution to the problem would be
ETS1.B: Developing Possible Solutions • Designs can be conveyed through sketches, drawings, or physical models. These representations are useful in communicating ideas for a problem's solutions to other people.	Sketched their solution to the problem and revised the sketch after testing the physical models
2-PS1.A: Structure and Properties of Matter • Different properties are suited to different purposes.	Selected and tested different building materials in the design of their solution
Crosscutting Concept	
Structure and Function	By building bridges and moving objects up and down a ramp, participated in two tasks to understand that structure and function help in solving certain problems

Source: NGSS Lead States 2013.

Connecting to the *Next Generation Science Standards*

The materials, lessons, and activities outlined in this chapter are just one step toward reaching the performance expectations listed below. Additional supporting materials, lessons, and activities will be required.

3-5-ETS Engineering Design www.nextgenscience.org/3-5ets-engineering-design	Connections to Classroom Activity
Performance Expectations	
3-5 ETS1-1: Define a simple design problem reflecting a need or a want that includes specified criteria for success and constraints on materials, time, or cost	Presented with and explained a design problem to include criteria and constraints
3-5 ETS 1-2: Generate and compare multiple possible solutions to a problem based on how well each is likely to meet the criteria and constraints of the problem	Examine multiple solutions to an assigned problem by creating individual designs, comparing those designs with their teammates, and selecting the best options

3-5-ETS Engineering Design www.nextgenscience.org/3-5ets-engineering-design	Connections to Classroom Activity
Performance Expectations	
3-5 ETS 1-3: Plan and carry out fair tests in which variables are controlled and failure points are considered to identify aspects of a model or prototype that can be improved	Constructed and tested their team prototype, modified the model after discussion, and competed in a water race with their model
Science and Engineering Practices	
Asking Questions and Defining Problems	Asked questions of each other and the teacher as they worked on solving a problem
Planning and Carrying Out Investigations	Designed, tested, and modified their boat
Constructing Explanations and Designing Solutions	Created two-dimensional sketches to serve as a model for their solution and explained their design to the class
Disciplinary Core Ideas	
ETS1.A: Defining and Delimiting Engineering Problems • Possible solutions to a problem are limited by available materials and resources (constraints). The success of a designed solution is determined by considering the desired features of a solution (criteria). Different proposals for solutions can be compared on the basis of how well each one meets the specified criteria for success or how well each takes the constraints into account.	Understood the criteria for and constraints given Compared their design with their teammates to select which features best meet the criteria for building a boat
ETS1.B: Developing Possible Solutions • At whatever stage, communicating with peers about proposed solutions is an important part of the design process, and shared ideas can lead to improved designs.	Had ongoing discussions with their teammates and the teacher associated with their boat design and modifications to be made after testing
ETS1.C: Optimizing the Design Solution • Different solutions need to be tested in order to determine which of them best solves the problem, given the criteria and the constraints.	Considered different designs to solve the problem both during the design feature and during the evaluation of the video boat races
3-PS2.A: Forces and Motion • Each force acts on one particular object and has both a strength and direction.	Identified that blowing through a straw creates a force that moves their boat across the bin of water
Crosscutting Concept	
Influence of Engineering, Technology, and Science on Society and the Natural World	Described their initial design and the changes they made to their boat prototype after testing it as well as why these changes were made

Source: NGSS Lead States 2013.

References

National Governors Association Center for Best Practices and Council of Chief State School Officers (NGAC and CCSSO). 2010. *Common core state standards.* Washington, DC: NGAC and CCSSO.

NGSS Lead States. 2013. *Next Generation Science Standards: For states, by states.* Washington, DC: National Academies Press. *www.nextgenscience.org/next-generation-science-standards.*

Resources

Beaty, A. 2007. *Iggy Peck, architect.* New York: Abrams Books for Young Readers.

Beaty, A. 2013. *Rosie Revere, engineer.* New York: Abrams Books for Young Readers.

Belloni, G. 2011. *Anything is possible.* Berkeley, CA: Owl Kids.

Novak, B. J. 2014. *The book with no pictures.* New York: Dial Books for Young Readers.

Van Dusen, C. 2005. *If I built a car.* New York: Puffin Books.

Yamada, K. 2013. *What do you do with an idea?* Seattle, WA: Compendium.

Internet Resources

Making Matters! How the Maker Movement Is Transforming Education *www.weareteachers.com/blogs/post/2015/04/03/how-the-maker-movement-is-transforming-education*

Middle School Cardboard Boat Race *www.youtube.com/watch?v=MHyqAaFKWe8*

STEM Design Challenge: Edible Cars *www.teachingchannel.org/videos/engineering-design-process-stem-lesson*

Why the Maker Movement Is Important to America's Future *http://time.com/104210/maker-faire-maker-movement*

NSTA Connection

Download the student data sheets and a list of additional resources at *www.nsta.org/SC1509.*

More Teaching Through Trade Books

Flying Machines

By Emily Morgan, Karen Ansberry, and Susan Craig

This chapter presents two fiction trade books you can use to inspire students to design and test various flying machines. In *Violet the Pilot*, Violet's mechanical genius inspires students in grades K–2 to test and compare different flying toys. Students in grades 3–5 read about the daring and crazy ideas of Captain Arsenio and his (mis)adventures in flight, then modify the design of a CD hovercraft to see how far they can make it go.

Trade Books

Violet the Pilot
By Steve Breen
Dial Books for Young
Readers, 2008
ISBN: 9780803731257
Grades K–2

Synopsis

While other girls play with dolls and tea sets, Violet plays with monkey wrenches and needle-nose pliers. The other kids make fun of her until she uses her mechanical genius to save the day!

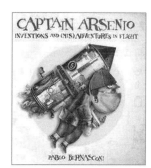

*Captain Arsenio:
Inventions and (Mis)
adventures
in Flight*
By Pablo Bernasconi
Houghton Mifflin,
2005
ISBN: 9780618507498
Grades 3–5

Synopsis

This amusing book features pages from the recently discovered diary of the fictional Captain Manuel J. Arsenio, in which he recorded his many failed (and humorous) attempts to create a flying machine.

Curricular Connections

The *Next Generation Science Standards* (*NGSS*) suggest that students not only learn about engineering but also have opportunities to engage in engineering practices. The *NGSS* break down the engineering design practices into three parts: defining and delimiting engineering problems, developing possible solutions, and optimizing design solutions. The basic ideas are the same, but they grow in complexity as students move to higher grade levels.

In this month's lessons, we have presented opportunities for students to engage in optimizing design solutions through testing and comparing flying machines. In the K–2 lesson, students compare the distance various flying toys can travel and learn that all flying machines have to work against the pull of gravity, which supports disciplinary core idea PS2.A: Forces and Motion, which states that "pushes and pulls can have different strengths and directions" (NGSS Lead States 2013). In the 3–5 lesson, students consider specific criteria and constraints while designing a model hovercraft, which supports the science and engineering practice of planning and carrying out investigations (NGSS Lead States 2013). Students also learn about the forces acting on the hovercraft and how they affect its motion, relating to disciplinary core idea PS2.A: Forces and Motion, which states that "each force acts on one particular object and has both a strength and a direction" (NGSS Lead States 2013).

Grades K–2: Testing Flying Machines

Purpose

Inspired by the book *Violet the Pilot*, students will test and compare three different flying machines.

Engage

Show students the cover of *Violet the Pilot* and ask them to infer from the picture and title what the book might be about. Invite them to look closely at the flying machine on the cover and identify some of the things it is made of (soap box, surfboard, bicycle tire, clown horn, and so on.) Read the story aloud, and be sure to point out the different flying machines that Violet made and the names she gave them, such as the Tub-bubbler, Pogo Plane, and Slide Glider.

Materials

The following materials are needed for this activity:

- Safety goggles (one for each student)
- Crayons
- File folder
- Journal inserts
- Tape

Each group needs the following materials:
- Three flying toys
  - Squeeze rocket
  - Hand helicopter
  - Foam glider
- Tape measure
- Three small sticky notes

Explore

Tell students that after Violet makes each new flying machine, she tests it. Like Violet, they are going to have a chance to test a flying machine. Give each student a small sticky note and have them draw a picture of Violet. They will be attaching one of these pictures to each of the flying machines they test. Give each student a Violet's Flying Machines student page (see NSTA Connection, p. 40). Have them draw and name their flying machine in the first box on the student page. Show students Violet's sketch on the inside cover of the book as an example of what a labeled drawing looks like. Next, create a starting line in the room and have students record how far that particular toy carried Violet. Students should wear safety goggles when using projectiles. Demonstrate how to measure with the tape measure. Explain it is important to do several tests or trials when they are collecting data, so they should record three trials and circle the longest distance of the trials on the student page.

Explain

Revisit page 16 of *Violet the Pilot* and point out the word "engineering." Ask students if they know what the author means by "Violet's engineering" (answers will vary). Tell students that you have a video from NASA that explains what engineering means and what engineers do. As they watch NASA for Kids: Intro to Engineering video (see NSTA Connection, p. 40), have them listen for what "engineering" means and what engineers do (*Common Core State Standards, English Language Arts [CCSS ELA]*, connection: Speaking and Listening, Comprehension and Collaboration; NGAC and CCSSO 2010). Then, have them share responses with the whole group. Ask, "How is Violet like an engineer?" Students should use examples from the text to answer the question. They could refer to any of the things that Violet builds and tests (*CCSS ELA* connection: Reading Literature, Key Ideas and Details; NGAC and CCSSO 2010). Remind students that the video explained that engineers design things to solve a problem. Then ask, "What problem was Violet working on?" Students should recognize that Violet was working on creating machines that would fly. Explain that gravity is the force that pulls things to the ground and that all flying machines have to work against the pull of gravity to fly. Ask them why they think Violet did so many test flights in the book. Students should remember from the video that engineers have to test what they build to know if it works well.

Elaborate

Tell students that, like Violet, they are going to test more flying machines. They will test a total of three different flying machines to see which is best at getting Violet safely from the start line to the finish line. Give each team a new flying toy, and have them stick a paper Violet to it. Students should follow the procedure in the explore section with the new toy, recording their results in the appropriate box on the student page. After all students have tested the second toy, give them the third toy to test, following the same procedure.

Evaluate

Have students fill out the last section of the student page, titled "The Winner." There they should identify the flying machine that took Violet the farthest distance and brainstorm ways to make that particular flying machine even better. For fun, you can create a flight journal out of a file folder for each student by opening a file folder and folding ⅓ of the bottom up horizontally to form a pocket. Then, fold it in half vertically, with the pocket on the outside. Next, fold it in half again to create four pockets (see photo). Students can then cut out the boxes on the student page and place one in each pocket. On the outside pocket, students should record one good thing (advantage) and one bad thing (disadvantage) of each flying machine. For example, for the glider, an advantage would be that it made a smooth landing and a disadvantage would be that it did not go as far as the squeeze rocket.

 Flying Machine # 1

What would you name this flying machine?

_____

Draw a picture of the flying machine below:

How far did Violet go in this flying machine?

Trial	Distance (cm)	Flight Observations
1		
2		
3		

Circle the longest distance for this flying machine.

A student writes in his flight journal.

Grades 3–5: Designing a Hovercraft

Purpose

After reading about the fictional Captain Arsenio and his (mis)adventures in flight, students will build and redesign a model hovercraft, given certain criteria and constraints.

Engage

Show students the cover of *Captain Arsenio: Inventions and (Mis)adventures in Flight*. Ask what they think the book might be about, and introduce the author and illustrator, Pablo Bernasconi. Explain that Bernasconi was only 16 years old when he got his pilot's license and he's been flying ever since. His love of strange machines and airplanes helped him imagine the eccentric and brave Captain Arsenio. Read the book aloud, stopping after reading page 7, "The Discovery." Ask students why an inventor would keep a diary. Explain that it is important to document each design and the results of every test to keep track of what worked and what didn't work. For each two-page spread introducing one of Captain Arsenio's six projects, hide the illustrations as you read the left-hand page. Have students close their eyes, listen, and imagine what each invention looks like. Then, reveal the illustration on the right-hand page. Discuss how their visualizations compare with the actual illustrations and how the illustrations contribute to what is being conveyed in the story (*CCSS ELA* connection: Reading Literature, Integration of Knowledge and Ideas; NGAC and CCSSO 2010). As you share the flight diary pages for each flying machine, be sure to point out all of the measurements and notes he records in the diary.

Materials

Each student needs the following materials:

- CD
- Mini marshmallow
- Permanent marker
- Balloon
- Sports bottle cap (caps can be saved from disposable water bottles)
- Goggles

Each team of two needs the following materials:

- Various supplies for modifying the hovercrafts (different-size and -shape balloons, foil, plastic wrap, other surfaces to attach to the bottom of the hovercrafts, and so on)
- Handheld air pump

Each class needs the following materials:

- Signs that say "Here" and "There"
- Hot glue gun (for teacher use only)

Explore

Tell students that you would like them to help Captain Arsenio test a seventh flying machine—the CD Hovercraft. Give each student a CD, a permanent marker, a marshmallow to represent Captain Arsenio, and a copy of the Captain Arsenio's Next Project student page (see NSTA Connection, p. 40). Have students follow the instructions on the student page to build and test the CD Hovercraft. Place hot glue on the bottom of the cap, and place the bottom opening of the cap over the hole in the middle of the CD. Be sure that glue does not cover the hole or leak onto the other side of the CD. Only the teacher should use the hot glue gun. Allow at least 10 minutes for the glue to cool and harden before students are allowed to touch it. (For specific instructions and photographs on building a CD hovercraft, see NSTA Connection, p. 40).

Explain

Have students share their results from their student pages. They should discover that the hovercraft moves farther when the cap is open. Explain that when the cap is open, the air rushes out of the balloon and creates a buffer between the ground and the bottom of the CD. This reduces the friction (rubbing) between the hovercraft and the ground, making the hovercraft able to move faster and farther. Show students a labeled diagram with arrows showing the forces acting on the hovercraft, including the following: the push from their hand (pointing in the direction the hovercraft moves), gravity (pointing down), friction and air resistance (both pointing in the opposite way the hovercraft moves). Discuss the direction of each force. Explain that gravity, air resistance, and friction all slow the motion of the hovercraft. Tell them that later they will have a chance to modify their hovercraft to move farther and will need to consider these forces when they begin designing.

Tell students that you have a video from Dragonfly TV (see Internet Resource, p. 40) that shows

---

Name _____

Hovercraft Designs – Part 1

Design a hovercraft that meets the following criteria:
1) Captain Arsenio is carried from "Here" to "There."
2) Captain Arsenio (the marshmallow) must stay aboard the hovercraft at all times.

Your design must be designed using the following constraints:
1) The hovercraft is powered by the air of only one balloon.
2) You must use only the supplies your teacher approves.

Brainstorm and sketch your ideas below.

Teacher Checkpoint

---

two students, Sara and Rachel, who built and tested a life-size hovercraft. The two girls design solutions for two problems they had with their original hovercraft design. Explain that a key step to designing an effective solution is to understand the problem. To fully understand the problem, you must understand the criteria and constraints involved in solving the problem. Explain that *criteria* are the conditions that must be met to solve the problem and *constraints* are the things that limit a solution, such as cost, availability of materials, safety, and so on. Tell them that as they watch the episode of Dragonfly TV titled "Hovercraft," you would like them to listen and watch for the criteria the girls were trying to meet when designing their hovercraft and the constraints they faced as they worked (*CCSS ELA* connection: Speaking and Listening, Comprehension and Collaboration; NGAC and CCSSO 2010). Students should recognize that in one of the problems Sara and Rachel solved, the hovercraft had to move over a rough surface such as concrete. In the other test, the hovercraft had to be steered through an obstacle course. Ask students what constraints the two girls dealt with as they worked on their hovercraft. Students should recognize that the girls had a limited amount of money to spend and could use only the supplies and tools they had at home. They also had limits on the areas and surfaces on which to test their hovercraft. See NSTA Connection (p. 40) for more videos and articles about hovercraft design.

Elaborate and Evaluate

Challenge students to pair up and design a hovercraft that will allow Captain Arsenio (a marshmallow) to go from a place designated in the room marked "Here" to another place in the classroom marked "There." Have students use the Hovercraft Designs—Part 1 student page to brainstorm ideas (see NSTA Connection, p. 40). Students can try different amounts of air, different-shape balloons, different surfaces on the bottom of the CD, and so on. Tell them you will need to approve their designs and materials before they begin building by signing or stamping the Teacher Checkpoint box. Then, students can create four different designs and test them. Give them a copy of the Hovercraft Designs—Part 2 student page to fill out for each of their designs. Their designs must meet the following two criteria and be made within the following two constraints:

Criteria

- Captain Arsenio is carried from "Here" to "There."
- Captain Arsenio (marshmallow) must stay aboard the hovercraft at all times.

Constraints

- The hovercraft is powered by the air of only one balloon.
- You must use only the supplies your teacher approves.

Have each student create a flight diary by opening a file folder and folding $1/3$ of the bottom up to a pocket. Then, have them fold it in half, with the pocket on the outside. Next, they should fold it in half again to create four pockets. Students can then cut out their student page boxes and place one in each pocket. On the outside of each pocket, students can write some comments that Captain Arsenio himself might say about his ride on each of their hovercrafts. On the back of the flight diary, have students draw a labeled diagram of the hovercraft, including arrows showing the direction in which each force (push from a hand, friction, gravity, and air resistance) is acting (*CCSS ELA* connection: Writing, Text Types and Purposes; NGAC and CCSSO 2010).

Connecting to the *Common Core State Standards*

This section provides the *Common Core State Standards (CCSS), English Language Arts (ELA),* and

CCSS Mathematics standards addressed in the activities to allow for cross-curricular planning and integration. The *CCSS* state that students should be able to do the following at grade level.

ELA

Speaking and Listening Standards for K–5: Comprehension and Collaboration

- Kindergarten: Confirm understanding of text read aloud or information presented orally or through other media by asking and answering questions about key details and requesting clarification if something is not understood.

- Grade 1: Ask and answer questions about key details in a text read aloud or information presented orally or through other media.

- Grade 2: Recount or describe key ideas or details from a text read aloud or information presented orally or through other media.

- Grade 3: Determine the main ideas and supporting details for a text read aloud or information presented in diverse media and formats, including visually, quantitatively, and orally.

- Grade 4: Paraphrase portions of a text read aloud or information presented in diverse media and formats, including visually, quantitatively, and orally.

- Grade 5: Summarize a written text read aloud or information presented in diverse media and formats, including visually, quantitatively, and orally.

Reading Standards for Literature K–2: Key Ideas and Details

- Kindergarten: With prompting and support, ask and answer questions about key details in a text.

- Grade 1: Ask and answer questions about key details in a text.

- Grade 2: Ask and answer such questions as *who, what, where, when, why,* and *how* to demonstrate understanding of key details in a text.

Reading Standards for Literature 3–5: Integration of Knowledge and Ideas

- Grade 3: Explain how specific aspects of a text's illustrations contribute to what is conveyed by the words in a story (e.g., create mood, emphasize aspects of a character or setting).

- Grade 4: Make connections between the text of a story or drama and a visual or oral presentation of the text, identifying where each version reflects specific descriptions and directions in the text.

- Grade 5: Analyze how visual and multimedia elements contribute to the meaning, tone, or beauty of a text.

The *CCSS* emphasize writing across all content areas, as seen by standard statement 10, which begins in grade 3 and states that students should "write routinely over extended time frames (time for research, reflection, and revision) and shorter time frames (a single sitting or a day or two) for a range of discipline-specific tasks, purposes, and audiences."

Furthermore, the *CCSS ELA* provide a standard related to the range of text types for K–5 where it indicates that students in K–5 should apply the reading standards to a wide range of texts, including informational science books.

Connecting to the *Next Generation Science Standards*

The materials, lessons, and activities outlined in this chapter are just one step toward reaching the performance expectations listed below. Additional supporting materials, lessons, and activities will be required.

K-2-ETS3 Engineering Design *www.nextgenscience.org/k-2ets1-engineering-design*	Connections to Classroom Activity
Performance Expectation	
K-2-ETS1-3: Analyze data from tests of two objects designed to solve the same problem to compare the strengths and weaknesses of how each performs.	Tested and compared different toy flying machines
Science and Engineering Practice	
Analyzing and Interpreting Data	Analyzed data about the test flights of three different flying machines and chose the most successful machine based on the data
Disciplinary Core Idea	
ETS1.C: Optimizing the Design Solution • Because there is always more than one possible solution to a problem, it is useful to compare and test designs.	Tested different designs of flying machines, compared structures of the machines, and compared the distances they fly
Crosscutting Concept	
Structure and Function	Identified advantages and disadvantages of the structure and shape of different toy flying machines that support reaching optimum distances

Source: NGSS Lead States 2013.

Connecting to the *Next Generation Science Standards*

The materials, lessons, and activities outlined in this chapter are just one step toward reaching the performance expectations listed below. Additional supporting materials, lessons, and activities will be required.

3-5-ETS1-2 Engineering Design *www.nextgenscience.org/3-5-ets1-2-engineering-design* **3-PS2 Motion and Forces** *www.nextgenscience.org/3ps2-motion-stability-forces-interactions*	Connections to Classroom Activity
Performance Expectation	
3-5-ETS1-2: Generate and compare multiple possible solutions to a problem based on how well each is likely to meet the criteria and constraints of the problem.	Identified the criteria and constraints in a video of children testing a real hovercraft, and then tested and compared their own model hovercraft designs
Science and Engineering Practices	
Developing and Using Models	Constructed a model hovercraft designed to travel the greatest distance

3-5-ETS1-2 Engineering Design *www.nextgenscience.org/3-5-ets1-2-engineering-design* **3-PS2 Motion and Forces** *www.nextgenscience.org/3ps2-motion-stability-forces-interactions*	**Connections to Classroom Activity**
Science and Engineering Practices	
Planning and Carrying Out Investigations	Planned and conducted investigations to test variables that affect the distance a homemade hovercraft can travel
Disciplinary Core Ideas	
ETS1.A: Defining and Delimiting Engineering Problems • Possible solutions to a problem are limited by available materials and resources (constraints).	Constructed a model hovercraft to meet the specified requirements within the constraints provided.
ETS1.B: Developing Possible Solutions • Tests are often designed to identify failure points or difficulties, which suggest the elements of a design that need to be improved.	Tested homemade hovercrafts and revised designs based on tests
PS2.A: Forces and Motion • Each force acts on one particular object and has both strength and a direction. An object at rest typically has multiple forces acting on it, but they add to give zero net force on the object. Forces that do not sum to zero can cause changes in the object's speed or direction of motion.	Observed the effects of forces on models and created labeled diagrams of hovercrafts using arrows to show the direction of the forces acting on the models
Crosscutting Concept	
Cause and Effect	Tested how changes in designs cause changes in the ability of the hovercrafts to meet the criteria and constraints presented

Source: NGSS Lead States 2013.

References

National Governors Association Center for Best Practices and Council of Chief State School Officers (NGAC and CCSSO). 2010. *Common core state standards.* Washington, DC: NGAC and CCSSO.

NGSS Lead States. 2013. *Next Generation Science Standards: For states, by states.* Washington, DC: National Academies Press. *www.nextgenscience.org/next-generation-science-standards.*

Internet Resource

Dragonfly TV: Hovercraft
   http://pbskids.org/dragonflytv/show/hovercraft.html

NSTA Connection

Visit *www.nsta.org/SC1310* for the student pages and additional resources and *bit.ly/YGHDAB* to see NASA for Kids: Intro to Engineering and NASA SCI Files: Hovercraft Development videos.

PART 2

GRADES PREK–2

PreK — K — 1 — 2

5

Gimme an *E*!

Seven Strategies for Supporting the *E* in Young Children's STEM Learning

By Cynthia Hoisington and Jeff Winokur

Early childhood educators have long debated how science should be introduced and taught to preschoolers. In the current STEM education climate, this conversation has expanded to include the role of engineering in the preschool curriculum.

Why is there increased interest in early childhood engineering? First, young children's constructive and dramatic play provides a natural context for identifying, addressing, and solving engineering design problems. As they create castles for people, corrals for horses, or garages for cars, children choose among available building materials and put them together in different ways with an eye to their structure's function ("Is this fence high enough to keep the horses in?"), strength ("Will the walls hold up the roof?"), and stability ("How can I keep the castle from tipping?"). Second, building experiences connect children to core concepts in physical science. Children observe the properties of the building materials they use (whether they are hard, soft, flexible, etc.) and experience the effects of applied and "natural" forces (including gravity and friction) acting on

the materials and their buildings. Their buildings stand, sway, or topple depending on how these properties and forces interact and on how children's structures are designed. Third, as the National Science Teachers Association position statement on early childhood science education makes clear (NSTA 2014), research shows that young children are capable of conceptual understanding in the STEM disciplines. From a young age, children generate ideas that help them make sense of the physical world and how it works (Duschl, Schweingruber, and Shouse 2007). Although their explanations are often scientifically incorrect ("You need to use tall blocks to make tall buildings."), they do stem from children's own reasoning about their prior observations and experiences with objects and materials. Young children can also develop the habits of mind that are integral to science and engineering, such as curiosity and persistence. And finally, *A Framework for K–12 Science Education* (NRC 2012) formally recognizes the close relationships among the STEM disciplines. It incorporates the idea that learning concepts and practices begin early and deepen over time

---

and across grade levels. All of these factors, taken together, strongly suggest that preschool-age children should be engaged in a range of developmentally appropriate and playful learning experiences in physical, life, and Earth sciences that center on key concepts or "big ideas." These experiences have the capacity to build a foundation for children's later understanding of the core ideas, crosscutting concepts, and science and engineering practices outlined in the *Next Generation Science Standards* (*NGSS*; NGSS Lead States 2013). It is important to emphasize that young children's STEM experiences should be expansive and not limited to the specific performance expectations for kindergarten and elementary grades.

A block area includes a variety of building materials.

Instructors and coaches in the professional development program Cultivating Young Scientists (CYS) worked with preschool teachers in mixed-age classrooms of three-, four-, and five-year-old students in Hartford, Connecticut, over five months as they implemented a unit on the topic of building structures (Chalufour and Worth 2004). In this chapter, we share the seven overlapping and mutually reinforcing strategies teachers used that effectively supported

children's learning in physical science and engineering.

Prepare the Environment for Investigating Structures

Preparing the environment means planning space, materials, and time for building explorations. CYS teachers arranged their existing "block areas" so that three to four children could build at one time and extended building into other learning areas. They collected a variety of building materials including wood, foam, and cardboard "blocks," intentionally incorporating different sizes, weights, shapes, and textures. Some teachers dug into closets, borrowed from other classrooms, scoured recycle centers, and collected common items (paper towel tubes, cardboard boxes) that could be transformed into building materials. Teachers integrated 20–40 minutes for building 2–3 times per week into their classroom schedules.

Teachers prepared to extend the unit to include ongoing explorations of the school and neighborhood buildings. They designated display spaces at children's eye level where photos, drawings, and descriptions of children's buildings would be posted as the unit progressed. In the meantime, they hung inspirational photos of houses, skyscrapers, and bridges, along with iconic structures such as the Eiffel Tower. When wall space was limited, teachers used the backs of shelf units or doors, or they placed binders with plastic sleeves in the block area. Paper, pencils, clipboards, and collage materials were collected for children's two-dimensional and three-dimensional building representations. Construction paper "blocks" provided additional representing options to younger children and children with limited fine-motor capacity. All of these opportunities enabled children to experience phenomena related to core building concepts. Figure 5.1 shows the relationship between the foundational concepts (at the center of the wheel) and opportunities for children to experience phenomena related to the concepts (outside of the wheel). Note that the impact of forces is less observable to children than the impact of materials and design on their structures.

FIGURE 5.1

Core Building Concepts and Structures Concepts

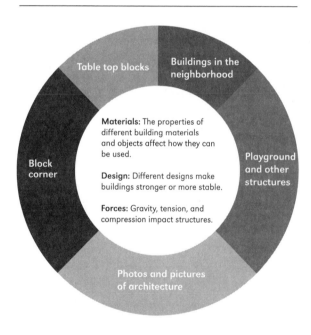

To ensure safety during the unit, teachers limited the number of children allowed in the block area, and they generated or reviewed existing rules with children, such as blocks are for building and builders can knock down only their own structures. CYS instructors asked teachers to reconsider safety rules that restricted the height of children's structures and suggested new strategies such as requiring builders to wear hard hats, allowing ample space for building and closely monitoring children's tall buildings.

Make Time for Teachers' Own Science and Engineering Investigations

To effectively plan for, facilitate, and assess children's learning in a building unit, preschool teachers need opportunities to participate in and reflect on their own collaborative building explorations. These experiences support their understanding of the relevant physical science concepts and immerse them in the practices essential to science and engineering (Wenglinsky and Silverstein 2006–2007). During CYS sessions, instructors facilitated teachers' inquiry-based explorations as teachers built tall towers, enclosures, and ramps and investigated and represented neighborhood buildings. Teachers noticed that using dense versus less dense materials and different sizes, shapes, and textures made a difference in the strength and stability of their structures. They discovered that the foundation was a critical design feature and that the need for stability imposed a constraint on how high they could build. Teachers also used science and engineering practices as they identified structural problems, drew and created models, measured their structures, debated about materials and design, identified patterns in what contributed to strong and stable structures, and generated ideas about successful building strategies. These experiences, along with discussions about how children learn in the content areas, familiarized teachers with how science and engineering practices might apply to children's building explorations.

A student creates an enclosure with cardboard blocks.

Sequence-Build Explorations Intentionally

Children build an understanding of core concepts over time and after many active experiences with related phenomena. CYS teachers intentionally sequenced children's ongoing building experiences from more *open* to more *focused* (Chalufour and Worth 2004). Initial open explorations over two or three weeks provided children with multiple opportunities to familiarize themselves with the building materials as they used them to build a variety of structures. Children noticed, for example, that cylinders sometimes rolled off their structures and that stacked wood blocks tended to slide. Early explorations leveled the playing field for children who had few prior experiences with

building. They also provided time for children and teachers to adjust to new routines, including representing and science talks, and provided teachers with preliminary data about their children's understanding and skills related to building.

Increasingly focused explorations reflected children's growing engineering skills, as well as their interests in building tall and in building homes for animals. Two towers investigations, lasting two to three weeks each, extended children's thinking about the properties of the building materials ("Which materials will build the tallest tower?") and focused children's attention on design ("How can we make our towers tall, but also strong and stable?"). A subsequent focused exploration of enclosures, stemming from children's interest in building animal homes, challenged children to create interior spaces, considering width and depth as well as height. Making walls, roofs, doors, and windows deepened students' thinking about properties ("Should I use heavy or lights blocks for the roof?") and design ("How can I add a door without the wall falling down?"). Enclosures explorations also emphasized the concept of form and

A teacher observes, facilitates, and documents children's building explorations during a conversation in the Explore stage.

function ("How does a bunny's home need to be made differently than a giraffe's home?").

Organize and Facilitate Children's "Minds-On" Building Explorations

Explicit frameworks help teachers organize children's science explorations and facilitate interactions that promote conceptual development and inquiry. CYS teachers used the Engage-Explore-Reflect (EER) cycle (Chalufour and Worth 2004) to ensure that each building exploration included multiple cycles of inquiry and a full range of minds-on as well as hands-on practices (see Figure 5.2).

FIGURE 5.2

Engage-Explore-Reflect Cycle

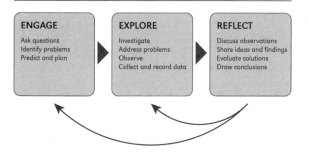

The EER cycle also enabled teachers to embed prompts—comments and questions that promote inquiry—within each phase of the cycle (Table 5.1, p. 48). During the Engage stage teachers used productive prompts that elicited children's prior knowledge about building structures and invited them to raise questions, identify problems, and make predictions. During the hands-on Explore stage, teachers encouraged children to observe their buildings and identify, address, and solve building challenges. Prompts for the Reflect stage helped children describe their building experiences using language, drawings, photos, and

demonstration and express their emerging ideas about how to make strong and stable structures. Those prompts were multi-functional and contributed to students' language development and assessment practices.

Integrate Opportunities for Language, Literacy, and Mathematics During Building Explorations

Science and engineering provide ideal contexts for language and literacy because communication is a critical aspect of both disciplines and because children are naturally motivated to communicate their observations, discoveries, and ideas. Building explorations addressed foundational skills in each of the four *Common Core State Standards* literacy strands: Language, Speaking and Listening, Writing, and Reading (NGAC and CCSSO 2010). Teachers facilitated 5–10 minute science talks using a small-group format, photos, and children's building representations to better support the participation of all children, including English language learners. Using productive prompts, teachers scaffolded children's ability to use language for asking questions, describing, making comparisons, and expressing conclusions. Throughout the unit, teachers introduced and emphasized increasingly challenging vocabulary words, including *build*, *blocks*, *structure*, and words that described the properties of the building materials. Teachers supported speaking and listening skills by encouraging children to maintain their focus on the topic, share their observations and ideas appropriately, and listen and respond to the contributions of other children.

Teachers fostered foundational writing and reading skills as they helped children create, share, and interpret representations. They invited children to use emergent writing to record data

TABLE 5.1

Examples of Productive Prompts

Activity	Engage	Explore	Reflect
Open Building Exploration	Let's explore the different materials we have for building. What do you notice about how these blocks look or feel? How are they the same or different? What would you like to build? Which blocks or other materials might you use?	Can you tell me about your building and how you are making it? Describe the different parts of your building. I notice that you are using the ___ blocks in this part and the ___ blocks in that part. How do the people, animals, or cars get into and out of the building?	Here are some photos I took of your structures. Would you like to describe your structures and how you made them? How well did those blocks work in your building or that part of your building? Why do you think they worked or didn't work so well? What was the easiest or hardest part of your structure to build?
Focused Towers Exploration 1	Let's look at photos of the tall structures you've been building. What do you notice about them? Which blocks do you think will work best for making really tall towers? Why do you think those blocks will be best? How could we find out?	Tell me about your tower and how you are building it. I notice that you are using the ___ blocks at the bottom or the top of your tower. Have you tried adding any other kinds of blocks to your tower? Here are some recording tools you can use to draw your tower.	Let's look at your drawings of your towers. Can you describe what you used to build your tower? Why did you decide to put the ___ blocks at the bottom or top? Which blocks do you think worked best for making really tall towers? Why do you think so?
Focused Towers Exploration 2	You made some really tall towers with the blocks and other materials. How might we make our towers stronger and more stable? How could we use this fan to find out how strong and stable the towers are?	Does your tower always fall down the same way? What parts stay standing? It looks like you are making your tower wider at the bottom or placing the blocks this way and then that way or putting the same numbers of blocks on each side. Let's draw and take photos of the towers.	Which of our towers were the strongest and most stable? How do you know? What do you notice about the towers that stayed up when we put the fan on them? What about the towers that didn't stay up? What are some different ways we found to make tall, strong, stable towers?

and modeled conventional writing as they transcribed children's dictation about their buildings. When teachers talked with children about their building drawings and stories and read them fiction and nonfiction books about building (including *How a House Is Built* [1996] by Gail Gibbons and *I Fall Down* [2004] by Vicki Cobb), they supported children's development of pre-reading skills.

Structures explorations enabled teachers to teach math concepts, language, and skills for a purpose. Teachers encouraged children to measure their towers using standard and non-standard measurement tools (their own bodies, unit blocks). They supported children's learning about spatial relationships ("How could you change the house so the teddy bear can stand up inside?") and patterns ("I notice that you placed foam, then wood, then foam, then wood. Why did you decide to use that pattern?").

Collect Assessment Data Related to Building From a Variety of Sources

Assessment in science is a continuous process of uncovering children's knowledge and skills in relation to the core concepts and science and engineering practices. The best assessment probes are embedded in the curriculum and promote, as well as assess, conceptual learning and inquiry (Snow and Van Hemel 2008).

CYS teachers collected assessment data in the context of children's building explorations and as they facilitated children's learning and inquiry during each phase of the Engage-Explore-Reflect cycle. As teachers interacted with children during the Explore phase, for example, they closely observed and recorded children's building behaviors. They also noted how the children approached and persisted at building, used

A documentation panel makes children's ideas and thinking explicit.

materials, designed their structures, and played and talked with each other. They made copies of children's building representations and transcribed what children said about them. During Engage-Explore-Reflect conversations, teachers noted how individual children communicated their building observations, experiences, and ideas. They could then individualize in the moment for children with a range of developmental levels, language skills, and social-emotional abilities by adding or removing materials, scaffolding language and vocabulary, or pairing students with a more knowledgeable peer, for example.

Reflect On, Document, and Use Data From Children's Building Explorations

When teachers reflect on and document data from children's explorations, they make children's thinking and learning visible. This process also serves to inform on-going planning. After the second towers exploration, CYS teachers collaboratively reflected on their building observations, photos, representations, and language samples. Each teacher created a documentation panel that illustrated what the teacher viewed as the most prominent aspects of his or her own children's learning up to that point. Individual panels highlighted, for example, children's abilities to investigate the properties of materials and design and the materials' impact on stability, identify and recreate patterns in their structures, and collect data about structures using a variety of measurement tools. The panels also illustrated children's thinking and emerging theories about building, such as blocks are needed on the sides to keep buildings balanced, wooden blocks make strong structures, and blocks must be placed carefully to make buildings stay up. Additionally, panels highlighted the playful, imaginative, and social nature of young children's authentic science and engineering explorations.

This collaborative reflection revealed that many children in classrooms were intentionally trying out and choosing different materials for different parts of their towers and developing tower-building strategies, such as creating a hard base to build on, placing blocks carefully with an eye to balance, and even widening and strengthening their towers' foundations. Some children were verbally sharing their ideas about how to build tall towers ("Wood blocks work better at the bottom because they're heavy and sturdy. The tower stays up better if you build on something hard instead of the rug."), indicating their readiness to address a second design challenge such as building enclosures for animals of different sizes. It also enabled teachers to identify children in their classrooms who would benefit from more explicit language supports, additional options for representing, and intentional grouping with peers during explorations and conversations. Additionally, teachers determined that some children would benefit from ongoing open explorations and individualized support for investigating, using, and observing different building materials.

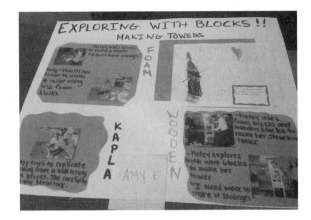

A documentation panel highlights one child's exploration of various building materials.

Teachers used their panels as the basis for follow-up conversations with children, further drawing out children's building interests and their ideas about building strong and stable structures. In doing so, teachers obtained assessment information that informed their planning of enclosures explorations. They also gained a deeper understanding of young children, how they think and learn, and the types of experiences and interactions that foster their learning in science and engineering.

Conclusion

Young children are curious and eager to engage in constructive and dramatic play by nature—but they must be taught to take advantage of these predispositions if they are to become more adept at thinking like scientists and engineers. Although preK performance expectations are not explicitly outlined in the *NGSS*, we have identified some of the ways in which young children's building experiences connect with and are foundational to developing specific practices, disciplinary core ideas, and crosscutting concepts (see "Connecting to the *Next Generation Science Standards*").

Consistent implementation of the teaching strategies described in this chapter requires that preschool teachers have time, training, administrative support, and a commitment to science and engineering education for young children. But the potential payoff is tremendous—an opportunity for preschoolers and their teachers to jump into the world of 21st-century STEM!

Connecting to the *Next Generation Science Standards*

The materials, lessons, and activities outlined in this chapter are intended for use in preK classrooms. Science experiences in preK by their nature are foundational and relate to later learning in K–12 classrooms. Because the *NGSS* performance expectations are for grades K–12, we have not included specific performance expectations but have identified the disciplinary core ideas that are addressed to show the link between these foundational experiences and students' later learning.

Science and Engineering Practices	*Connections to Classroom Activity*
Asking Questions and Defining Problems	Identified a challenge or problem (built a tall tower or a home for animals)
Developing and Using Models	Developed ideas and tested which designs work best for their specific purposes
Constructing Explanations and Designing Solutions	Developed ideas about which materials work best under different circumstances
Disciplinary Core Ideas	*Connections to Classroom Activity*
ETS1.A: Defining and Delimiting Engineering Problems • Asking questions, making observations, and gathering information are helpful in thinking about problems	Investigated a range of building materials and designs and gathered information about the benefits and challenges of each
ETS1.B: Developing Possible Solutions • The ability to build and use physical models is an essential part of translating a design into a finished product.	Continued to work on and improve their designs Used building materials in multiple ways based on their properties
PS1.A: Structure and Properties of Matter • Matter can be described and classified by its observable properties. A great variety of objects can be built up from a small set of pieces.	Completed cycles of building, knocking down, and rebuilding structures with blocks
Crosscutting Concepts	*Connections to Classroom Activity*
Structure and Function	Designed structures (e.g., towers) that met a challenge or served a purpose
Stability and Change	Explored the way various designs affect stability

Source: NGSS Lead States 2013.

References

Chalufour, I., and K. Worth. 2004. *Building structures with young children*. St. Paul, MN: Redleaf Press.

Cobb, V. 2004. *I fall down*. New York: HarperCollins.

Duschl, R., H. Schweingruber, and A. Shouse, eds. 2007. *Taking science to school: Learning and teaching science in grades K–8*. Washington, DC: National Academies Press.

Gibbons, G. 1996. *How a house is built*. New York: Holiday House.

National Governors Association Center for Best Practices and Council of Chief State School Officers (NGAC and CCSSO). 2010. *Common core state standards*. Washington, DC: NGAC and CCSSO.

National Research Council (NRC). 2012. *A framework for K–12 science education: Practices, crosscutting concepts, and core ideas*. Washington, DC: National Academies Press.

National Science Teachers Association (NSTA). 2014. National Science Teachers Association Position statement: Early childhood science education. *www.nsta.org/about/positions/earlychildhood.aspx*.

NGSS Lead States. 2013. *Next Generation Science Standards: For states, by states*. Washington, DC: National Academies Press. *www.nextgenscience.org/next-generation-science-standards*.

Snow, C. E., and S. B. Van Hemel, eds. 2008. *Early childhood assessment: Why, what, and how*. Washington, DC: National Academies Press.

Wenglinsky, H., and S. C. Silverstein. 2006–2007. The science training teachers need. *Educational Leadership* 64 (4): 24–29.

NSTA Connection

Visit *www.nsta.org/SC1509* for a list of resources.

The EDP-5E

A Rethinking of the 5E Replaces Exploration With Engineering Design

By Pamela Lottero-Perdue, Sonja Bolotin, Ruth Benyameen, Erin Brock, and Ellen Metzger

Many preservice and practicing elementary teachers are familiar with the 5E learning cycle (Bybee 1997). This cycle provides a relatively simple, alliteratively memorable framework for teaching science in which lessons (or even entire units of instruction) consist of five distinct phases: Engagement, Exploration, Explanation, Elaboration/Extension (hereafter, Extension), and Evaluation (Bybee et al. 2006; Bybee 2015).

With the inclusion of engineering in the *Next Generation Science Standards*, some may wonder if the 5E learning cycle is still a reasonable approach to structuring a lesson in which children engage in engineering design. We argue that the engineering design process (EDP) *can be* situated within the 5E framework, with the most significant changes being: (a) replacement of the Exploration phase with an EDP phase and (b) consideration of the different evaluation roles of students and teachers (Figure 6.1). In this chapter, we share our experiences introducing preschoolers (preK) and kindergarteners (K) to engineering through an hour-long lesson about designing structures.

FIGURE 6.1

Engineering Design Process 5E, or EDP-5E

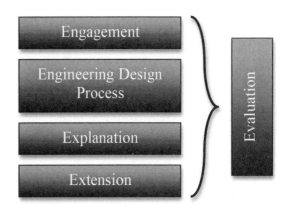

The young children we feature in this chapter were students at Tunbridge Public Charter School in Baltimore, Maryland, where preK and K classrooms learn thematic units of instruction (e.g., dinosaurs or plants) that connect science and other subject matter learning. At this school, students in grades 1–5 learn science year-round and also learn one Engineering is Elementary unit of instruction per year (e.g., EiE 2011a). Before Ms.

Lottero-Perdue and her student teacher interns in her early childhood methods class visited, the preK and K teachers at Tunbridge—coauthors Ms. Ruth Benyameen, Ms. Erin Brock, and Ms. Ellen Metzger—had not yet taught engineering to their students. The brief lesson featured in this chapter represented the beginning of the preK and K students' engineering education journeys, which would continue into their later elementary years.

In the lesson, a toy stuffed crab needed a tower upon which to sit and then a house in which to live, and—taking each problem in turn—the children-as-engineers designed a solution for each. A fun activity within the purposefully silly context of a crab tower and crab house, this lesson did the important work of attending to performance expectations, practices, crosscutting concepts, and disciplinary core ideas related to science and engineering (see "Connecting to the *Next Generation Science Standards*," p. 59).

In what follows, we explain the engineering design process 5E (or, EDP-5E) modified-5E lesson format; feature the creative contributions and questioning strategies of one of the student teacher interns, Ms. Bolotin, as she led the lesson in a K classroom; and describe and depict how the students were engaged in and responded to the lesson. We feature Ms. Bolotin and her students to illustrate how the lesson was enacted across all three classrooms.

The Engineering Design Process

The EDP that we used for this lesson was based on the Engineering is Elementary EDP (EiE 2011a). In this EDP, students working in teams begin by being presented with a problem. Before problem solving, students must consider constraints (i.e., limitations), criteria (i.e., factors that define what a successful designed solution will be able to do), and relevant knowledge they have that could be useful when solving the

problem (e.g., scientific knowledge). Students then brainstorm possible solutions to the problem and—after selecting one possible solution to try—plan, create, test, and evaluate the success of that possible solution. Students then repeat these actions, creating a second designed solution to improve or optimize their first possible designed solution. If time warrants, additional ones can be developed. Typically, designed solutions are shared and described by teams so they can see how others made decisions and how other teams' designs performed.

Engagement

One of the engagements used across the preK and K classrooms was an interactive read-aloud of the book *Iggy Peck, Architect* (Beaty and Roberts 2007). This book is about a little boy named Iggy Peck who loves to build towers, bridges, or buildings with materials available to him in his world, be it in his house, at the playground, or classroom. His talent comes in handy when his second-grade class is stranded on an island during a picnic, and he employs his skills of using available resources to build a bridge.

Ms. Bolotin posed many questions to her students both during and after the read-aloud to elicit from them key ideas in the text and important engineering practices, including "What does Iggy Peck like to do?" *(He likes to build things.)* "What does Iggy Peck use when he builds those things?" *(Anything he can find!)* "What did Iggy Peck do when his class got stranded?" *(He started gathering materials to use to build a bridge.)* "Did Iggy Peck just start putting materials together for the bridge, or did he plan his bridge first?" *(He drew his idea for a bridge first using a stick to draw in the dirt.)* "Was Iggy Peck successful? Did he rescue his class?" *(Yes!)* Ms. Bolotin concluded her read-aloud by suggesting that Iggy Peck was a budding engineer and that today, they too would be engineers like Iggy Peck!

We should note here that other books could be used to engage students in thinking about engineering or structures, including books that feature characters in groups underrepresented within the engineering field. For example, the author and illustrator of *Iggy Peck, Architect* recently published *Rosie Revere, Engineer* (Beaty and Roberts 2013) in which Rosie invents, tries to solve problems, and dreams of becoming an engineer. *The Best Beekeeper of Lalibela: A Tale from Africa* (Kessler and Jenkins 2006) is a story of an African girl who designs and redesigns ways to keep a honeybee hive vibrant and producing honey, despite initial strong skepticism by male villagers that she could do so. Yet another option for a text-based engagement is Christy Hale's *Dreaming Up: A Celebration of Building* (2012), which takes readers on trips to see towers and other structures in Malaysia, Egypt, and China while featuring diverse children building model structures.

Engineering Design Process (Exploration in 5E)

The children were shown two toy stuffed animals—a crab and an alligator—and an image of water drawn on paper lying on the floor. "This little crab has a big problem," exclaimed Ms. Bolotin. "The very hungry alligator wants to eat the crab, but the crab does not want to be eaten! Our goal is to build a tower for the crab on the floor that will be tall to keep the alligator far away and strong to support the crab. Can you help?"

Ms. Bolotin introduced the constraints by sharing that each team could only use 40 5-oz. paper cups and 12 sheets of construction paper to build their towers and that they could not cut or bend the cups or papers. (Tip: Doing so enables teachers to reuse supplies!) She also reminded them about

the criteria: their towers should be as tall as possible and strong enough to hold the crab by itself.

Ms. Bolotin organized students into teams of two, and teams began to brainstorm their ideas of how the cups and papers could be stacked and oriented. Teams drew their ideas on paper, just as Iggy Peck had done in the dirt, and then picked an idea to try first. Before handing out the cups and papers, Ms. Bolotin paused to discuss safety, asking that teams be mindful of where their friends were building so as not to trip over their tower construction sites in the room, potentially injuring themselves, their friends, or their friends' towers. She also reminded students to be respectful of their teammates and to share in the process of tower construction. Teams began the process of building, ever mindful of the hungry alligator, and were able to use their frightened crab to test their towers as they constructed (Figure 6.2).

FIGURE 6.2

Team Building Tower With Cups and Construction Paper

It was then time to test! While all the teams watched with excitement, Ms. Bolotin asked a team member to carefully place the crab on top of his or her tower to see if the tower could hold the crab. Most teams were successful in holding up the crab; however, some collapsed upon bearing the crab's weight.

Before giving students a chance to tear down their structures and try again, Ms. Bolotin engaged them in a critical reflection of their first tower. This enabled students to analyze and interpret their testing data, consider cause (i.e., their design) and effect (i.e., their testing results), and consider the relationship between structure and function in their designs. Ms. Bolotin asked, "Was your tower high enough to protect the crab? Could it be higher? Was your tower strong enough to hold the crab? How can you improve the tower?" After considering the answers to each question, teams planned, created, and tested a second tower. With few exceptions, this second tower was higher or stronger than the first, indeed optimizing and improving their first designed solutions (see Figure 6.3).

Explanation

At the end of the second test, Ms. Bolotin asked the children to look around at other teams' designs. She posed the following general question: "When you looked around, what did you notice?" More specifically, she asked: "Do all of the towers support the crab? Do all of the towers protect the crab by keeping the crab high and away from the alligator? Do all the towers look alike? How are they different?" The students noticed that some structures were taller than others, some used more cups in between layers of paper than others, and some stacked cups on top of one another. Although they acknowledged that some towers were not as tall or strong, they often noted other attributes of those towers (e.g., "It was a good idea to try to stack cups!").

Furthermore, students were asked to reflect on how their first and second towers compared to

FIGURE 6.3

One Group's Second, Improved Tower

one another. Ms. Bolotin asked them to consider, "What did you do to try to make your second tower taller or stronger than the first? Was your first or second design more successful? How do you know?" Children were encouraged to articulate reasons for their attempted improvements in their second towers, but—in the few cases when second towers failed—were also told that sometimes engineers have to try many, many times to improve their ideas.

Extension

The aim of the extension was to reinforce the big ideas discussed in the Explanation phase (e.g., that different designed solutions can solve the same problem and that a second try is a valuable engineering strategy). The problem was changed, yet the now-familiar materials and material constraints (40 cups, 12 sheets of paper) remained the same. "Now," said Ms. Bolotin, "we need to be engineers again. We want to create a house for our crab." She explained that this would be like a house or apartment where we might live, not a place where real crabs might live. She elicited from students the basic features of such a place—for example, having at minimum a roof, floor, walls, and door.

Ms. Bolotin elicited the criteria from students for house features through questioning, by asking, for example: "How will we know if a house that we create is a good one for this crab? How tall should the house be? How wide should we make the door? What should the walls and roof look or be like?" Students' responses included that the door should be wide enough for the crab to enter and exit, the floor space should be wide enough for the crab to move around, the walls should be "solid," and the roof should not sag or have holes in it.

Teams again brainstormed, planned, created, and tested their designs (Figure 6.4). The testing process entailed moving the crab into, around, and out of the house, determining if criteria were met. When teams were finished with their first design, they were encouraged to look at other teams' designs and then go back to their own and improve or embellish it somehow. Typically, these improvements consisted of making wider doorways, adding swinging doors, adding windows, or filling in gaps in a wall. Ms. Bolotin posed similar questions that she posed in the explanation section of the 5E lesson, having students reflect on their own and others' design processes.

Evaluation

There are three ways that evaluation occurred for this EDP-5E lesson. First, it was the teams who were charged with evaluating their towers or houses against established or elicited criteria. This self-evaluation was critical and in keeping with the work of engineers who must

FIGURE 6.4

Testing the Crab House

self-evaluate long before clients and others evaluate their work. Ms. Bolotin did not assign her own summative judgment to the performance of the teams' designs; she did not need to, and doing so would have overly emphasized performance over process.

Second, Ms. Bolotin continually monitored students as they created their towers and houses. She checked for understanding of the problem, constraints, and criteria, and she guided and helped as necessary. Formative assessment

FIGURE 6.5

Side View of a Child's Tower

began with Ms. Bolotin's questioning of students' understanding of the text in the Engagement phase. It continued as she assessed students' responses to questions about their towers in the Explanation phase and as students responded to similar questions upon reflecting on their houses in the Extension phase.

Third, Ms. Bolotin asked students to draw either their best house or best tower and to explain to her what problem it solved and how it worked to solve the problem. Figure 6.5 is a child's side-view of his tower, depicting the cups-paper-cups-paper (and so on) layering strategy. Figure 6.6 represents a top view of another child's tower, demonstrating how her cups (circles) are distributed between the papers, holding up both the corners and center of the tower.

In addition, we developed a simple rubric to assess students' engagement in the EDP. This rubric is available online (see NSTA Connection, p. 60). Criteria on this rubric include understanding and application of problem, constraints, and criteria; sharing and listening to ideas in a team; sharing materials and resources; planning; reasoning regarding designed plans, comparing first and second designed solutions, and comparing designed solutions across teams.

Conclusion

The 5E lesson framework remains a useful tool within science education. Replacing the science-based Exploration phase with a complete EDP is one way to re-engineer this tool. Aside from this E-shift, ideas about the other Es are worthy of note. The job of the Engagement phase in the EDP-5E is to pique students' interest in the problem or problem context. There are myriad ways to do this, however, beyond reading a book. There are also many ways to extend learning in the Extension phase. Either as an Engagement or an Extension, students could draw or take pictures of

FIGURE 6.6

Top View of Another Child's Tower

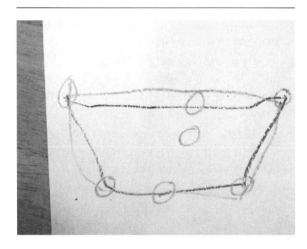

structures in their world, take apart a toy or model structure to see how it is constructed (i.e., reverse engineer the toy or model), or look at images of structures like towers and houses online.

Finally, although not emphasized in this introductory engineering activity, relevant science concepts may be included throughout the EDP-5E. For example, they may be addressed in the Engagement phase to remind students about prior relevant science knowledge. Science concepts may be highly relevant during the EDP section as students are making design decisions affected by science. For instance, next year in one of their first-grade units, *Thinking Inside the Box: Designing Plant Packages*, the current K students will try to design a package that can keep a plant healthy (EiE 2011b). Also, during the Explanation phase, teachers can prompt students to explain how they considered science ideas (e.g., plants need air and water) in their engineering design solutions.

The EDP-5E represents a way to situate engineering design—perhaps a relatively novel process for early childhood and elementary teachers—into the familiar framework of the 5E to honor both of these methods of learning and doing. The preservice and practicing teachers involved in this lesson were able to see the students' deep engagement, enthusiasm, and desire to problem solve. With just a few exceptions, the crabs were safe and sheltered, thanks to the child engineers!

Connecting to the *Next Generation Science Standards*

The materials, lessons, and activities outlined in this chapter are just one step toward reaching the performance expectations listed below. Additional supporting materials, lessons, and activities will be required.

K-2-ETS1 Engineering Design *www.nextgenscience.org/K-2ets1-engineering-design*	Connections to Classroom Activity
Performance Expectations	
K-2-ETS1-2: Develop a simple sketch, drawing, or physical model to illustrate how the shape of an object helps it function as needed to solve a given problem.	Drew models to communicate their learning
3-5-ETS1-3: Analyze data from tests of two objects designed to solve the same problem to compare the strengths and weaknesses of how each performs.	Used the engineering design process (EDP) to design, create, test, compare, and optimize crab towers and houses
Science and Engineering Practices	
Planning and Carrying Out Investigations	Carried out the EDP by engineering designed solutions
Analyzing and Interpreting Data	Analyzed and interpreted data from testing results
Designing Solutions	Analyzed, evaluated, and revised their own and others' designs based on testing results
Obtaining, Evaluating, and Communicating Information	Collaborated in teams as they engineered and communicated designs within and across teams
Disciplinary Core Ideas	
ETS1.B: Developing Possible Solutions • Designs can be conveyed through sketches, drawings, or physical models. These representations are useful in communicating ideas for a problem's solutions to other people.	Used the EDP to develop possible solutions to the crab tower and house problems
ETS1.C: Optimizing the Design Solution • Because there is always more than one possible solution to a problem, it is useful to compare and test designs.	Cycled through the EDP, which includes considering each problem, constraints, and criteria; brainstorming with peers; and planning, creating, testing, and optimizing the crab tower and house solutions
PS1.A Structure and Properties of Matter • A great variety of objects can be built up from a small set of pieces.	Created towers and houses using simple paper cups and pieces of construction paper

K-2-ETS1 Engineering Design www.nextgenscience.org/K-2ets1-engineering-design	Connections to Classroom Activity
Crosscutting Concept	
Structure and Function	Engineered a tower using construction paper and paper cups and learned that certain orientations and combinations of these materials affect the tower's strength and stability
	Engineered a crab house where they use these materials to meet specific needs and functions of a house (e.g., roof, floor, walls, and door)

Source: NGSS Lead States 2013.

References

Beaty, A., and D. Roberts. 2007. *Iggy Peck, architect.* New York: Abrams.

Beaty, A., and D. Roberts. 2013. *Rosie Revere, engineer.* New York: Abrams.

Bybee, R. W. 1997. *Achieving scientific literacy: From purposes to practices.* Portsmouth, NH: Heinemann.

Bybee, R. W. 2015. *The BSCS 5E Instructional Model: Creating teachable moments.* Arlington, VA: NSTA Press.

Bybee, R. W., J. Taylor, A. Gardner, P. Van Scotter, J. Powell, A. Westbrook, and N. Landes. 2006. *The BSCS 5E Instructional Model: Origins, effectiveness, and applications.* Colorado Springs, CO: BSCS.

Engineering is Elementary (EiE). 2011a. The engineering design process. *In Engineering is Elementary: About EiE.* Boston, MA: Museum of Science. *www.eie.org/overview/engineering-design-process.*

Engineering is Elementary (EiE). 2011b. *Thinking inside the box: Designing plant packages.* Boston, MA: Museum of Science. *www.eie.org/eie-curriculum-units/thinking-inside-the-box-designing-plant-packages.*

Hale, C. 2012. *Dreaming up: A celebration of building.* New York: Lee and Low Books.

Kessler, C., and L. Jenkins. 2006. *The best beekeeper of Lalibela: A tale from Africa.* New York: Holiday House.

NGSS Lead States. 2013. *Next Generation Science Standards: For states, by states.* Washington, DC: National Academies Press. *www.nextgenscience.org/next-generation-science-standards.*

NSTA Connection

Visit www.nsta.org/SC1509 for a list of resources.

7

Can a Student Really Do What Engineers Do?

Teaching Second Graders About Properties Using Filter Design Within a 5E Learning Cycle

By Sherri Brown, Channa Newman, Kelley Dearing-Smith, and Stephanie Smith

As an elementary science teacher, you may be questioning if students in second grade can actually do engineering practices. You may also be questioning if you have the expertise needed to teach engineering practices. Our response to both questions is a definitive yes. We—a science teacher educator, a water company informal educator, and a second-grade teacher—developed and co-taught a three- to four-day instructional unit within the authentic context of water filtration.

In designing the unit, we considered the fact that most elementary teachers have approximately 30–45 minutes daily for a science lesson. After reviewing *A Framework for K–12 Science Education* (*Framework*; NRC 2012), the *Next Generation Science Standards* (*NGSS*; NGSS Lead States 2013), and several lessons focused on water quality (Dacko and Higdon 2004; Moyer and Everett 2011; Walker, Kremer, and Schlüter 2007), we designed this unit to support instruction on properties of solids and liquids, specifically *NGSS* 2-PS1.A: Structure and Properties of Matter. During the unit, students collaborate to test different materials to determine which materials have the properties that are best

suited for filtering soil particles from water. We followed the 5E lesson plan model (Bybee et al. 2006) to organize the entire unit.

A Framework for Engineering

The *Framework* states that "children are natural engineers … they spontaneously build sand castles, dollhouses, and hamster enclosures and use a variety of tools and materials for their own playful purposes" (NRC 2012, p. 70). The *NGSS* also supports engineering design in the K–2 grades, where students are "introduced to problems as situations that people want to change. They use tools and materials to solve simple problems, use different representations to convey solutions and compare different solutions to a problem, and determine which is best" (NGSS Lead States 2013, Appendix I, p. 105). Thus, it is imperative that engineering practices be introduced in early grades so that students have opportunities to develop creative, critical-thinking processes to solve problems.

Because the *Framework* recommends that science teachers provide experiences for students to see "how science and engineering pertain to real-world problems and to explore opportunities to apply their scientific knowledge to engineering design problems once this linkage is made" (NRC 2012, p. 32), we contextualized an investigative science unit that allows students to explore properties of materials while using engineering practices (EPs; see Figure 7.1). Collaborations between the University of Louisville science teacher educator and the Louisville Water Company informal educators have occurred over a decade; our prior co-teaching experiences included several summer science camps for seventh graders and K–12 science teachers. The behind-the-scenes tours of various water company facilities have provided middle school students, K–12 science teachers, and the science teacher educator a literal glimpse into the realities of how our water is cleaned and distributed to our community. Without this collaboration, the authenticity of the lesson would have been compromised.

Engage

We introduced the filter unit by showing a cartoon movie clip where Nemo the fish clogs the filter (see Internet Resources, p. 68). After the video clip, we asked the class what Nemo clogged; then we stated we were going to explore filters for the next few days and document our exploration (e.g., predictions, observations, data, models, and conclusions) in a science notebook. To determine students' prior knowledge of filters, we asked them to identify a filter by circling items that contained a filter on a Where's the Filter? activity sheet (see Figure 7.2). To support English language learners, we converted the English terminology (e.g., *clothes dryer*, *aquarium*, *tea bag*, *swimming pool*, and so on) to students' native languages (see Internet Resources, p. 68, for free online dictionaries). From a cursory review of students' sheets, we noticed that most of the

FIGURE 7.1

Essential Practices of Science and Engineering

1. Asking questions (for science) and defining problems (for engineering)

2. Developing and using models

3. Planning and carrying out investigations

4. Analyzing and interpreting data

5. Using mathematics and computational thinking

6. Constructing explanations (for science) and designing solutions (for engineering)

7. Engaging in argument from evidence

8. Obtaining, evaluating, and communicating information

Source: NRC 2012, p. 29.

students were unaware that the water treatment plant and the tea bag used a filter. After students had completed the sheet, we asked them to share their reasons for selecting certain items. We then guided the students' discussion to conclude that actually *all* of the items on the activity sheet contain a filter. Introduce other filters as appropriate, such as health care masks and air conditioner filters. Last, we pointed out that because of the infiltration (pun intended!) of technology in our lives, the term *filter* has other frequent uses (e.g., cameras filter light, e-mail systems filter SPAM, headphones filter background noise, sunglasses filter light, and colanders filter pasta).

At this point, the class discussed their working definition of *filter* and decided it was something that takes out or removes solid particles (e.g., soil) from liquids. From this definition, we asked the class if they could design a filter that removed soil from water. Specifically, the class defined

FIGURE 7.2

Activity Sheet: Where's the Filter?

7-6   Section 7: The Fabulous Filter

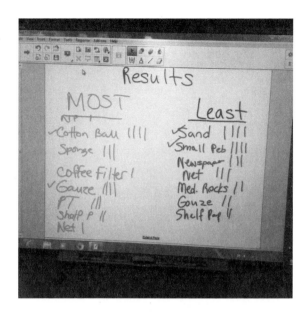

Tally marks indicating the type of filter used by each student group that removed the most and least amount of "dirt" from the water

our problem (EP1) as "What type of materials are needed in our filter to remove soil from a water solution?" We avoided "clean water" language because this was not a feasible outcome from our materials. The activity sheet and class *filter* definition were students' first science notebook entry. To effectively use science notebooks in the elementary classroom, we have implemented strategies from *Nurturing Inquiry: Real Science for the Elementary Classroom* (Pearce 1999).

Explore

We began our exploration by asking students to be careful and avoid placing anything in their mouths throughout this lesson; even water should be treated cautiously as a material in science class. Students must wear splash goggles

when working with *any* liquid. We introduced the test materials visually (see Figure 7.3, p. 64, for a materials list) and explained that we would later use some of these materials to construct a filter model (EP2). For English language learners in the classroom, we labeled materials in English and their native language. Using their prior experiences with each of the 11 materials, the students individually made predictions as to which material would remove the most soil from water and which would remove the least. For example, students with fishing experiences thought a net would filter best.

In preparation for the lesson, we constructed five pitchers of "dirty" water (15 ml commercial topsoil per 1 L of water) and placed one pitcher atop a piece of white laminated paper on a tray. We provided each student a Predict and Test (Figure 7.3, p. 64) activity sheet and goggles, and each two- or three-member student group was provided with 11 clear plastic cups and 11 red-marked clear plastic cups with holes in bottom. (In advance, we used a nail to

FIGURE 7.3

Activity Sheet: Predict and Test

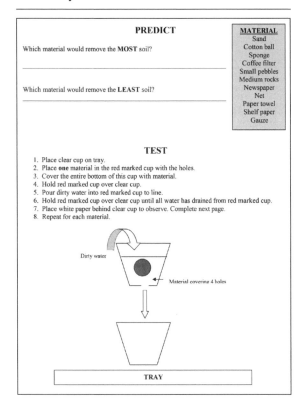

<table>
<tr><td colspan="2" align="center">PREDICT</td><td align="center">MATERIAL</td></tr>
</table>

PREDICT

Which material would remove the **MOST** soil?

_____

Which material would remove the **LEAST** soil?

_____

MATERIAL
Sand
Cotton ball
Sponge
Coffee filter
Small pebbles
Medium rocks
Newspaper
Net
Paper towel
Shelf paper
Gauze

TEST
1. Place clear cup on tray.
2. Place **one** material in the red marked cup with the holes.
3. Cover the entire bottom of this cup with material.
4. Hold red marked cup over clear cup.
5. Pour dirty water into red marked cup to line.
6. Hold red marked cup over clear cup until all water has drained from red marked cup.
7. Place white paper behind clear cup to observe. Complete next page.
8. Repeat for each material.

Dirty water

Material covering 4 holes

TRAY

puncture three to five holes in the bottom of each cup.) We placed containers of the 11 test materials, basins for waste, and paper towels at four supply stations. Before testing, the teacher educator modeled how student groups should test each material without actually pouring the dirty water into the cup. The teacher educator also demonstrated how to place the plain white paper behind the clear cup and how to record observations. Working in their groups, students tested the materials (EP3) and recorded their observations in written text (CCSS.ELA-LITERACY.W.2.7; NGAC and CCSSO 2010) and technical drawings. After students completed their testing, they reviewed and interpreted their data as a team (EP4) to answer four sentence starters on a Claim and Evidence data

sheet (see Figure 7.4). Students stated which two materials removed the most amount of soil and which two materials removed the least amount of soil. After completing their activity and data sheets, students glued them into their science notebooks as entry 2. We reviewed the sheets to tally students' test data to determine the amount of materials needed for the next day.

FIGURE 7.4

Data Sheet: Claim and Evidence

The two materials that removed the MOST soil were _____ and _____.

My reasons are _____ _____.

The two materials that removed the LEAST soil were _____ and _____.

My reasons are _____ _____.

Explain

According to a review of students' Claim and Evidence sheet, we reported the class tally in a pre-constructed table on the interactive white-board for the most effective and least effective material to remove soil from water. The class data set of 19 student responses were organized as frequency tally marks by each tested item under two columns, "removed the most soil" and "removed the least soil." We provided multiple opportunities for students to interpret the tabular data (EP5) (*Common Core State Standards, Mathematics,* Measurement and Data; NGAC and CCSSO 2010); for example, we asked them what material the class selected most often as being most effective in removing soil from water (i.e., cotton ball and gauze each had four responses).

We then asked several groups to share their reasons for selecting materials that worked best and that worked least to remove the soil. Students provided their reasons for their claims (EP4). During their sharing, we guided students' discussion to conclude that some materials worked better to filter the bigger stuff, while some worked better to filter the smaller stuff. The class implemented a vocabulary term from their previous explorations of solids and liquids by saying that some materials have "properties" that are better suited for an intended purpose than others (i.e., sand did not work well because it absorbed too much of the soiled water; gauze worked well because it had spaces to let water, but not soil, through). We revisited and used prior learning in discussing that all of the tested filter materials were solids; solids have a definite shape unless we physically alter it. However, a liquid does not have a definite shape; it takes the shape of the container (i.e., water takes the shape of the cup it is in). At this point, a discussion of sand ensued; however, after observing closely, students found that sand is made of small solid particles or grains you can rub between your fingers. We explained that engineers use models to analyze current and proposed systems, determine flaws, test possible solutions, and communicate designs to others (NRC 2012). Thus, in "becoming" an engineer, students designed, constructed, and tested (and potentially re-designed) a filter on the basis of their previous test data (EP6).

Elaborate

Students used their material test data to design their filter on the Water Filter Design activity sheet (see NSTA Connection, p. 69). Continuing as teams, students selected three materials and the order of those materials that would make up their filter. After we reviewed and approved their designs, students constructed their filter from a provided 2 L precut container (remove bottle's bottom with scissors and place gauze with rubber band over bottle's spout). The Water Filter Design activity sheet was science notebook entry 3. Although we did not have a restriction on how much of the filtering material they were able to use, students did not ask or attempt to make more than one single layer with the material.

Before beginning the Elaboration phase, we prepared the dirty water materials (e.g., trays, pitchers, and white paper) and each station (e.g., materials, basins for waste, and paper towels). As student groups held their 2 L bottle filter and cup, we poured 100 ml of dirty water into each filter. After water drained from the filter, students reflected and wrote a summary of their filter's effectiveness based on evidence (EP7 and EP8).

After students were seated on the carpet, the water treatment informal educator discussed each filter. She asked students observational questions, such as, "Do you believe soil particles remain in the cup?" "Did the filter remove soil?" and "What material worked well?" She removed materials from each filter and walked around the carpet to show students that each material did remove some soil from the water; however, some materials were better than others. The class concluded that each filter did work at some level to remove the soil. If time allows, students could re-construct their filter based on class discussion and their answer to the question, "If we could make our filter again, we would …." To include quantitative measurements to this lesson, one could add the constraints of time required for filtration or the amount of material needed. If cost of each material is used, then one would add a monetary function of the filter's (i.e., the system's) total cost.

Evaluate

Formative evaluation was ongoing throughout all phases of this unit learning cycle. The engagement phase provided a cursory view of students' previous "filter" content knowledge. During the exploratory phase, we interacted and questioned students as they tested and recorded data for each material. During the explain phase, we gained information from students' explanation of why certain materials were better than others in filtering the soil from the water. Within the elaboration phase, we evaluated students' application of their learning to the filter design. The evaluation summary (see NSTA Connection, p. 69) provided evidence of students' use of claims with supportive evidence.

Conclusion

We summarized our water filter unit learning by making connections to the the Influence of Engineering, Technology, and Science on Society and the Natural World standard, specifically that "every human-made product is designed by applying some knowledge of the natural world and is built using materials derived from the natural world" (NRC 2012, p. 213). Since we had a sink available in the classroom, we turned on the water to visually display the previously referenced clean water from the water treatment plant. We asked students, "How does a water treatment plant clean our water?" Students guessed "filter." Students then viewed the *Teachers' Domain Water Treatment Plant Video* (see Internet Resources, p. 68); some vocabulary in the video may need clarification, such

as reservoir, impurities, particles, chemicals, and residential. The water company informal educator provided various sealed water samples for students to view at their desks; samples included water from the surface of our local river, water from the bottom of our local river, and processed water from the company. She also displayed a large graduated cylinder model of a water company filter; the water company's actual filter beds are 9 ft. deep containing layers of coal, sand and different size rocks (larger rocks are on the bottom). She explained that these Earth materials naturally filter water over a long period of time (e.g., tens, hundreds, thousands of years). Last, she showed students laminated photographs of the filtering rooms from the local water treatment facility.

In closing, we encourage teachers to connect with their local water company educational personnel or informal educators for a personal water company tour, student-group tour, or class visit. Locally, our community's citizens have access to Crescent Hill Reservoir and Historic Gatehouse (see Internet Resources, p. 68). Our city's water company informal educators offer educational tours of the Gatehouse, and anyone visiting the reservoir site has access to tour the entire reservoir. After constructing filters, viewing online resources, and collaborating with water company informal educators, any teacher should be able to connect his or her lessons to water treatment facility processes and engineering practices.

Connecting to the *Next Generation Science Standards*

The materials, lessons, and activities outlined in this chapter are just one step toward reaching the performance expectations listed below. Additional supporting materials, lessons, and activities will be required.

2-PS1 Matter and Its Interactions *www.nextgenscience.org/2ps1-matter-interactions* **K-2-ETS1 Engineering Design** *www.nextgenscience.org/k-2ets1-engineering-design*	**Connections to Classroom Activity**
Performance Expectations	
2-PS1-2: Analyze data obtained from testing different materials to determine which materials have the properties that are best suited for an intended purpose	Tested materials to determine which two filtered the most and which two filtered the least soil from water based on the materials' properties
K-2-ETS1-1: Ask questions, make observations, and gather information about a situation people want to change to define a simple problem that can be solved through the development of a new or improved object or tool	Asked questions, made predictions, tested materials, and gathered information (technical drawings and descriptions) to determine which materials would filter the most and least soil from water
K-2-ETS1-3: Analyze data from tests of two objects designed to solve the same problem to compare the strengths and weaknesses of how each performs	Deconstructed and discussed filter designs as a class and concluded that even though some filter designs were better than others, each filter worked at some level to remove the soil
Science and Engineering Practices	
Asking Questions and Defining Problems	Defined problem as "What type of materials are needed in our filter to remove soil from a water solution?"
Developing and Using Models	Designed and tested a 2 L filter model and compared all student models
Planning and Carrying Out Investigations	Tested materials and recorded observations in written text and technical drawings
Analyzing and Interpreting Data	Reviewed and interpreted data as a team to support claims of which materials removed the least and most amount of soil followed by interpreting class evidence to determine effectiveness
Obtaining, Evaluating, and Communicating Information	Wrote summary that included evidence collected during investigation of filter materials
Disciplinary Core Ideas	
PS1.A: Structure and Properties of Matter • Different kinds of matter exist and many of them can be either solid or liquid, depending on temperature. Matter can be described and classified by its observable properties.	Determined which solid materials have the properties that are best suited for filtering soil particles from water

2-PS1 Matter and Its Interactions www.nextgenscience.org/2ps1-matter-interactions **K-2-ETS1 Engineering Design** www.nextgenscience.org/k-2ets1-engineering-design	**Connections to Classroom Activity**
Disciplinary Core Ideas	
K-2-ETS1.A: Defining and Delimiting Engineering Solutions • A situation that people want to change or create can be approached as a problem to be solved through engineering. • Before beginning to design a solution, it is important to clearly understand the problem.	Solved the given problem through engineering Asked questions, made observations, and gathered information concerning the problem
K-2-ETS1.C: Optimizing the Design Solution • Because there is always more than one possible solution to a problem, it is useful to compare and test designs.	Evaluated several solutions and determined that components of many worked to solve the problem
Crosscutting Concept	
Cause and Effect	Conducted simple test to gather evidence to support or refute material selection and order of for 2 L filter model

Source: NGSS Lead States 2013.

References

Bybee, R. W., J. A. Taylor, A. Gardner, P. Van Scotter, J. Carlson, A. Westbrook, and N. Landes. 2006. *The BSCS 5E Instructional Model: Origins and effectiveness.* Colorado Springs, CO: BSCS.

Dacko, M., and R. Higdon. 2004. Inquiring about water quality. *Science Scope* 41 (9): 34–36.

Moyer, R., and S. Everett. 2011. Charcoal—Can it corral chlorine? *Science Scope* 35 (2): 10–15.

National Governors Association Center for Best Practices and Council of Chief State School Officers (NGAC and CCSSO). 2010. *Common core state standards.* Washington, DC: NGAC and CCSSO.

National Research Council (NRC). 2012. *A framework for K–12 science education: Practices, crosscutting concepts, and core ideas.* Washington, DC: National Academies Press.

NGSS Lead States. 2013. *Next Generation Science Standards: For states, by states.* Washington, DC: National Academies Press. *www.nextgenscience.org/next-generation-science-standards.*

Pearce, C. R. 1999. *Nurturing inquiry: Real science for the elementary classroom.* Portsmouth, NH: Heinemann.

Walker, M., A. Kremer, and K. Schlüter. 2007. The dirty water challenge: A water filtration activity helps upper elementary students develop their investigation skills. *Science and Children* 44 (9): 26–29.

Internet Resources

Dictionary Boss
www.freedict.com/onldict/spa.html

Google Translate
http://translate.google.com

Louisville's Crescent Hill Filtration Plant, Reservoir and Gatehouse
www.louisvillewater.com/newsroom/crescent-hill-reservoir-and-gatehouse

Nemo™ Clogs Filter
www.youtube.com/watch?v=B1HmDc8zDw0

PBS Learning Media Earth Water Filter
*www.pbslearningmedia.org/resource/ess05.sci.ess.
earthsys.waterfilter/earth-water-filter*

Teachers' Domain Water Treatment Plant Video
*www.teachersdomain.org/resource/ess05.sci.ess.watcyc.
h2otreatment*

The Fabulous Filter
*http://tappersfunzone.com/sites/tapperfunzone.com/files/
handbooks/the_fabulous_filter.pdf*

NSTA Connection

**Visit *www.nsta.org/SC1407* for the activity
sheets and the evaluation summary.**

Catch Me if You Can!

A STEM Activity for Kindergarteners Is Integrated Into the Curriculum

By Kimberly Lott, Mark Wallin, Deborah Roghaar, and Tyson Price

Studies have shown that the number of students entering the science, technology, engineering, and math disciplines in college is dropping (National Academy of Sciences 2010). In recent years, STEM-related activities have become more common in schools with the hope of generating greater student interest in STEM. *A Framework for K–12 Science Education* (NRC 2012) spelled out the importance of teaching engineering processes alongside scientific inquiry, which paved the way for the strong emphasis on engineering in the *Next Generation Science Standards* (*NGSS*; NGSS Lead States 2013).

A STEM activity is any activity that integrates the use of science, technology, engineering, and mathematics to solve a problem. Traditionally, STEM activities are highly engaging and may involve competition among student teams. For example, many elementary schools are involved in building the strongest bridge out of popsicle sticks, the longest slinging catapult, or the most protective structure for an egg so it can be dropped from high heights. Although students are highly engaged in these types of STEM activities, the activities are usually isolated from regular classroom instruction. A more effective use of STEM activities is to integrate them into classroom lessons. This approach allows the teacher to create a STEM unit in which students are actively learning science content and then applying that knowledge to design a solution to a perceived problem.

Young children are natural engineers and often spontaneously build elaborate structures while they are playing (NRC 2012). For this reason, the early elementary years are often the most appropriate time to begin to incorporate deliberate engineering design problems. The following chapter presents a STEM unit that was completed over two weeks in a kindergarten classroom. The unit focused on the problem of building a trap to catch a mischievous gingerbread man (see NSTA Connection, p. 78, for objectives). Some might worry that the use of anthropomorphism might be misleading to young children. However, anthropomorphic texts have been shown to provide meaningful scientific learning (Gomez-Zwiep and Straits 2006). Although this problem was fictitious, the unit incorporated all aspects of STEM, and students learned how those aspects can be integrated to solve real problems encountered in their daily lives.

Science

To introduce this STEM unit, the teacher gathered the students at the rug for story time. The story for that day was the traditional folktale *The Gingerbread Man*. The teacher read the story up to the part where the gingerbread man ran away from the third or fourth animal (right before he meets the fox) and then closed the book. She then told the students about several "events" that had happened recently at their school. For example, books were knocked off of a library shelf, dirty footprints were all over the lunchroom tables, some papers were knocked off the principal's desk, and some of the classroom crayons were broken in half. "I think the gingerbread man is loose in our school," the teacher told them. "What do you think we can do?" "We could stay up all night in the school and see if we could see him," one student suggested. "We could make a trap!" another student excitedly exclaimed. "That's a great idea," several children agreed.

The children were now faced with a problem. To solve this problem, students first needed to explore the science that surrounds the problem. Before this STEM unit, the students had completed a physical science unit where they had discovered that objects have different properties (e.g., size, shape, texture, weight, and color). They had also explored the movement of non-living things (e.g., fast, slow, zigzag, up and down, back and forth) and how the properties of materials can affect their movement.

This unit connects to the *NGSS* performance expectation K-PS2 Motion and Stability: Forces and Interactions, which states that students will "plan and conduct an investigation to compare the effects of different strengths or different directions of pushes and pulls on the motion of an object" (NGSS Lead States 2013). The unit focuses on disciplinary core idea PS2.A. Forces and Motion (NGSS Lead States 2013). Along with the science standard, this unit also addresses standard K-2-ETS1: Engineering Design and meets the performance expectations K-2-ETS1-1, K-2-ETS1-2, and K-2-ETS1-3 (NGSS Lead States 2013). This unit also incorporates the following science and engineering practices: Defining Problems, Planning and Carrying Out Investigations, Analyzing and Interpreting Data, and Using Mathematics and Computational Thinking. It also supports the crosscutting concept of Cause and Effect.

Designing Solutions

The perceived need created by the gingerbread man problem made for an engaging learning environment. The students first made drawings of how they would trap the gingerbread man. This was useful preassessment for the teacher because it indicated which students had prior knowledge with traps and how they worked. The students then shared their drawings with a partner and students were encouraged to make changes to their drawings based on conversations with their peers.

After making their initial drawings, students were given several materials to explore, including cardboard, fabric, rubber band, marble, and cork. Students were asked to consider the properties of these materials and which would be most useful for making a gingerbread man trap. They recorded their answers on the student sheet (see NSTA Connection, p. 78). The students concluded some materials would be most useful because they are strong. "Your trap has to be able to hold the gingerbread man once you catch it," one student remarked. Other materials, such as fabric and string, were also considered useful because the gingerbread man might get "tangled up" and therefore trapped.

Working in pairs, the students were then given cardboard, tape, scissors, rubber bands, pipe cleaners, paper cups, and yarn. Using those materials, students practiced building traps, paying particular attention to the pushes and pulls of the materials that caused the "springing" of the trap. The students discovered that when making their traps something was always balanced (either a trap door, a dangling cup from a string as a "net," or a propped-up box). A push or pull was then used to cause this object to become unbalanced and "spring" the trap. "All he has to do is bump this stick and bam—the box will fall on him," one student remarked. The teacher used a checklist to evaluate the students' proposed traps (see NSTA Connection, p. 78).

Technology and Engineering

Technology is the product of engineering. This product can be created during an engineering activity or students may use it during either science or engineering activities. So how can students use what they learned about properties of materials and forces to design a gingerbread man trap? While gathered at the rug, the teacher presented the students with the following materials: a poster board, a wire coat hanger (straightened), string, cups, and tape. The teacher made sure students sat at a safe distance from the straightened wire hanger so that they would not come in contact with its ends during the demonstration. The teacher asked, "How could we use the poster board in our trap? What might the string and cups be used for?" The teacher then bent the coat hanger wire in the shape of a *C* and asked, "How might this be used?

The teacher asked, "How will the gingerbread man spring the trap?" The teacher then demonstrated by attaching the string to the cup, running the string over the coat hanger wire, and pulling the string on the other side of the wire hanger so that the cup stayed suspended in air. The teacher asked, "What would happen if I let go of the string?" Students predicted that the cup would

fall. The teacher let go of the string and the cup fell down. The students quickly discovered that they needed something to hold the string in order for the gingerbread man to spring the trap. Through these types of guided questions, the teacher and students came up with a basic trap design (see Figure 8.1). To keep the data collection manageable for this age of student, the teacher wanted to keep the designs fairly consistent, so only limited variables could be manipulated. For this reason, she worked with the students in their initial trap designs through guided inquiry. This scaffolding is essential to the learning progression of these developing engineering skills.

The teacher then introduced the word *technology* to the students. She explained that technology is when you use science and engineering to

FIGURE 8.1

Basic Trap Design Instructions

1. Cut poster board into strips about 12 in. x 22 in. Bend poster across width at about the 6 in. mark of the length to create a base for the trap.
2. Bend hanger into *C* shape and create a loop for string at top.
3. Tape hanger onto poster board using strong tape.
4. Attach string (4a) to the bottom of a clear cup (4b). Punch holes in the poster board at each *X* and feed the string through as shown in the diagram. Then, place a bead at the opposite end of the string.

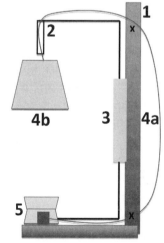

5. Glue two small clear cups together end to end to form a bait holder. Place bead under bait holder to set the trap.

8

FIGURE 8.2

Picture of a Student Trap

develop a tool that can then be used in our daily lives. These tools are sometimes called *inventions*. She then read several poems from *Incredible Inventions* by Lee Bennett Hopkins (2009). Those poems were used to illustrate that technologies are found everywhere, from Band-Aids to roller coasters, and that they can be developed by anyone, even young boys and girls. *The Boy Who Harnessed the Wind: Young Readers Edition* by William Kamkwamba and Bryan Mealer (2012) can also be read to further illustrate the point that technology and engineering solutions can be developed by anyone, regardless of age and cultural background. After reading the poems, she asked the students for other examples of technology and explained that the students had developed a new technology when they used the science they had learned about materials and forces to make their traps.

Engineering and Math

After the class had developed their trap design, the teacher then explained that they were going to be "acting like engineers" and wrote the new word on the board. The teacher explained that an engineer uses technology to design a solution to a problem and read the story, *Engineering the ABC's: How Engineers Shape Our World* by Patty

O'Brien Novak (2009). Throughout the following engineering design and testing process, the teachers assessed the students through observation and questioning using a rubric (see NSTA Connection, p. 78).

The teacher then asked, "How can we get the gingerbread man to want to come into our trap?" "It needs to look inviting like a gingerbread house," one student hypothesized. Another student thought, "It needs some type of bait." The teacher explained that when engineers are making their designs, sometimes they have to make multiple changes to find the perfect solution. The teacher wrote *inviting* and *bait* on the board and explained that they would need to work on them one at a time to see which design feature was most successful.

The students started with the appearance of their traps to make them more inviting. Before this class, the teacher and student teacher made one basic trap design for each pair of students. Teachers and parent volunteers also helped the students with the final assembly and decoration of their traps. Working in pairs, they colored gingerbread house pictures along with Disney characters and added them to their traps to entice the gingerbread man to spring the trap. Students also added cotton balls to the cups (trap) so it would look like a cloud to further fool that tricky gingerbread man. Once students were finished with the outside appearances of their traps, they set them in the halls all around the school building to see if they could catch the gingerbread man (see Figure 8.2).

The following day, the teacher created a bar graph on the board. She explained that engineers collect data and make graphs to tell them if their designs are working. She and the students then went out around the school to observe the traps to see if any had caught the gingerbread man. As the students entered back into the classroom, each pair of students indicated with an X if they had "caught" or "not caught" the gingerbread man and then gathered back on the rug. The students were somewhat disappointed that no one had caught the gingerbread man. The teacher reminded the students that engineers often have to make changes to their designs to finally find a design that works.

Referring to the graph, the teacher asked, "Does making our traps inviting seem to matter?" Becuase no one had caught the gingerbread man, the students concluded that the appearance of the traps was not enough to attract the gingerbread man. "Now let's try adding bait," the teacher suggested. She offered three types of bait for the students to choose from: jewels, gold coins, and crystals. Working in their pair groups, the students selected one type of bait and added it to their traps.

The following day, the teacher put a graph on the board. Each pair of students was given a sticky note based on the bait in their trap. Again, they observed their traps and indicated on the graph if their trap had made a catch. Students were more excited today because some of the traps had made a "catch." Before school, the teacher had placed a gingerbread man cookie (in a plastic bag) in the traps that contained the jewels. Referring to the graph again, the teacher asked, "Does the bait in your traps seem to matter?" "Yes!" the students exclaimed. "Would anyone like to change your bait?" the teacher asked. A resounding "Yes!" was received from the groups that had not made a catch.

The students who had not made a catch changed their bait to jewels, and all students reset their traps. The teacher placed two gingerbread man cookies in the traps this time so each student would have one to eat. The students then collected data again and created a third bar graph on the board. Referring to the last graph, "So were your traps successful?" asked the teacher. "The gingerbread man must like jewels!" exclaimed one student. The teacher

explained to the students the engineering processes that they had just completed. They had used engineering to design a solution to a problem. They used mathematical data to evaluate their designs and then made revisions based on this data (see NSTA Connection, p. 78, for the rubric). The teacher also pointed out that they all had been engineers.

Conclusion

At the conclusion of this unit, the teacher read a letter left to the class by the gingerbread man:

Dear Mrs. Roghaar's Class,

I am leaving your school to continue on my way. I have a fox to meet and a river to cross, but I wanted to tell you that your traps worked! I was able to escape, but since you were such clever students to trap me, I left you some gingerbread cookies to enjoy.

Have a great day!

The Gingerbread Man

While the students enjoyed eating the gingerbread cookies, the teacher reminded them of the science, technology, engineering, and math that they had used to solve the school's gingerbread man problem.

This STEM activity was not an isolated activity but an ongoing unit in this kindergarten classroom where students were introduced to how STEM can be used to solve real-world problems. Not all students were familiar with the ideas of technology or engineering, but after completing this unit they noticed examples of technologies in their daily lives and actively participated in the engineering process. During this unit, students started to become aware that even though science, technology, engineering, and math are different, they are intricately related in the problem-solving process.

Connecting to the *Next Generation Science Standards*

The materials, lessons, and activities outlined in this chapter are just one step toward reaching the performance expectations listed below. Additional supporting materials, lessons, and activities will be required.

K-PS2 Motion and Stability: Forces and Interactions www.nextgenscience.org/kps2-motion-stability-forces-interactions **K-2-ETS1 Engineering Design** www.nextgenscience.org/k-2ets1-engineering-design	**Connections to Classroom Activity**
Performance Expectations	
K-PS2-1: Plan and conduct an investigation to compare the effects of different strengths or different directions of pushes and pulls on the motion of an object	Given various materials, planned and built different traps, paying particular attention to the pushes or pulls of the materials that caused the trap to spring
K-2-ETS1-1: Ask questions, make observations, and gather information about a situation people want to change to define a simple problem that can be solved through the development of a new or improved object or tool	Decided that to solve the gingerbread man problem a trap would need to be developed
K-2-ETS1-2: Develop a simple sketch, drawing, or physical model to illustrate how the shape of an object helps it function as needed to solve a given problem	Sketched potential gingerbread man traps and developed physical working models based on initial sketches
K-2-ETS1-3: Analyze data from tests of two objects designed to solve the same problem to compare the strengths and weaknesses of how each performs	Graphed results from traps to determine which trap was most effective in catching the gingerbread man
Science and Engineering Practices	
Defining Problems	Defined the problem of catching the gingerbread man
Planning and Carrying Out Investigations	Manipulated different materials to determine which might be most effective for the functioning of the traps
Designing Solutions	Developed and tested different trap designs
Analyzing and Interpreting Data	Gathered evidence and constructed graphs to analyze data
Using Mathematics and Computational Thinking	Analyzed the number of successful and unsuccessful traps to determine their effectiveness
Disciplinary Core Ideas	
PS2.A: Forces and Motion • Pushing or pulling on an object can change the speed or direction of its motion and can start or stop it.	Investigated forces and resulting motion and balance of trap door, dangling cup, or propped up box as well as push or pull used to cause objects to become unbalanced and then move to spring the trap

K-PS2 Motion and Stability: Forces and Interactions www.nextgenscience.org/kps2-motion-stability-forces-interactions **K-2-ETS1 Engineering Design** www.nextgenscience.org/k-2ets1-engineering-design	**Connections to Classroom Activity**
Disciplinary Core Ideas	
ETS1.A: Defining and Delimiting Engineering Problems • A situation that people want to change or create can be approached as a problem to be solved through engineering.	Identified the situation to be changed
ETS1.B: Developing Possible Solutions • Designs can be conveyed through sketches, drawings, or physical models. These representations are useful in communicating ideas for a problem's solutions to other people.	Sketched traps, built working models, recorded information concerning the success of traps, and communicated results
ETS1.C: Optimizing the Design Solution • Because there is always more than one possible solution to a problem, it is useful to compare and test designs.	Identified the variables in trap designs and design success or failure and considered options for improving the design based on data
Crosscutting Concept	
Cause and Effect	Manipulated and observed pushes or pulls and how they spring the traps Observed and recorded that when bait was added to the trap the gingerbread man would sometimes get caught and that certain types of bait were more effective than others

Source: NGSS Lead States 2013.

References

Gomez-Zwiep, S., and W. Straits. 2006. Analyzing anthropomorphisms. *Science and Children* 44 (3): 26–29.

Hopkins, L. B. 2009. *Incredible inventions*. New York: Greenwillow Books.

Kamkwamba, W., and B. Mealer. 2012. *The boy who harnessed the wind: Young readers edition*. New York: Dial Books for Young Readers.

National Academy of Sciences. 2010. *Rising above the gathering storm, revisited: Rapidly approaching category 5*. Washington, DC: National Academies Press.

National Research Council (NRC). 2012. *A framework for K–12 science education: Practices, crosscutting concepts, and core ideas*. Washington, DC: National Academies Press.

NGSS Lead States. 2013. *Next Generation Science Standards: For states, by states*. Washington, DC: National Academies Press. *www.nextgenscience.org/next-generation-science-standards*.

Internet Resource

Edith Bowen Laboratory School: STEaM
https://sites.google.com/a/littleblue.usu.edu/stem/home

Trade Books

Llewellyn, C. 2008. *Exploring materials*. North Mankato, MN: Sea to Sea Publications.

Mason, A. 2005. *Move it: Motion, forces and you*. Toronto, ON: Kids Can Press.

Murphy, S. J. 2004. *Tally O'Malley*. New York: HarperCollins.

Novak, P. O. 2009. *Engineering the ABC's: How engineers shape our world*. Northville, MI: Ferne Press.

Rivera, E., and R. Rivera. 2010. *Rocks, jeans, and busy machines: An engineering kids storybook*. San Antonio, TX: Rivera Engineering.

NSTA Connection

Visit *www.nsta.org/SC1312* for the STEM objectives, student pages, and rubrics.

9

Inviting Engineering Into the Science Lab

Two Guided-Inquiry Lessons Display the Integration of *NGSS* Engineering Practices Into Traditional Elementary Labs

By Seth Marie Westfall

Whether you have created your own content-based science labs or you use a published science program, the onset of the *Next Generation Science Standards (NGSS)* require us as teachers to look at what we do with a slightly different lens. For example, the science program my district has adopted has several components that repeat for each lesson. One of those components is a science lab or lab activity. In most cases, this activity requires the teacher to walk students through a series of steps with the purpose of illustrating the particular concept being taught. The students are being told what, how, and in what order to perform those steps. According to Zion and Mendelovici (2012), this kind of activity falls under the realm of structured inquiry. They define this type of inquiry as an activity in which "students investigate a teacher-presented question through a prescribed procedure, and receive explicit step-by-step guidelines at each stage, leading to a predetermined outcome" (p. 384).

I have been teaching science this way since the beginning of my teaching career. However, after some recent work with the *NGSS*, I have had to rethink my use of the lab activity that

comes with my program. The new standards were designed, in part, to prepare students for college and careers with a strong base in "critical thinking and inquiry-based problem solving" (NGSS Lead States 2013, p. xv). The *NGSS* have a set of engineering standards for each grade level. As stated in the NGSS Executive Summary, these standards are written with the intent of getting students at all age levels to apply engineering practices throughout science content.

The set of steps that have been laid out in my program's lab activity do not allow very much room for students to ask questions such as "Why?" "What if?" or "How?" There are no problems for them to solve. Zion and Mendelovici (2012) would consider this kind of questioning to fall under the term *guided inquiry*. They define this type of inquiry as an opportunity for students to "investigate questions and procedures that teachers present to them, but the students themselves, working collaboratively, decide the processes to be followed and the solutions to be targeted" (p. 348).

I have included examples of two lessons I have adapted from my second-grade program to incorporate the science and engineering practices into my lab activities through the use of guided inquiry.

Solids, Liquids, and Gases

To work toward meeting the science and engineering standards, I have had to make a few modifications. During my unit on solids, liquids, and gases, the students are required to make a homemade freezer to illustrate the changing of matter (juice) from a liquid to a solid (ice pop) as part of their lab experience. The program directs me to guide them step by step in creating this freezer, including telling students what materials to use, how to combine them to make the freezer, and the purpose of each material. This activity seemed a logical place to incorporate the engineering practices because the kids were building something. Other opportunities might be areas of lessons in which the students are expected to ask or answer questions, test possible solutions to or identify problems, develop or improve a tool or object, or create a model to explain events or concepts. So, rather than using the structured-inquiry approach dictated in the manual, I decided to modify the activity to incorporate more of a guided-inquiry approach. See NSTA Connection (p. 86) for a more detailed lesson plan.

The original activity dictated a step-by-step guideline depicting how to use a bucket, ice, newspaper, and salt to create a homemade freezer. Instead, I gathered several examples of homemade ice cream makers. I informed my students that they were going to use these ice cream makers to make assumptions about how the devices were designed and the purpose behind those design components. What was the purpose for each part and why was it put there? They worked in groups to reverse-engineer each ice cream maker. At first, students were nervous and unsure. They had never been expected to do something like this before. Students made comments about how hard

this would be or that it was impossible. Once I put them in their groups and they started examining the ice cream makers, they began talking about what they looked like. They became highly engaged and began sharing some of their background knowledge and experiences with making ice cream. At first, the students were confused about where the ingredients and ice would go. Some of the preliminary guesses were that the ice would go in the metal cylinder in the middle. Others guessed that the ice, salt, and ingredients all got mixed together in the metal cylinder. Once students began sharing their background knowledge and experiences, they began to see that the ingredients would mix in the metal cylinder and the ice would go in the space between the cylinder and the outer wall. Only a couple of students were able to share that the salt gets mixed in with the ice. None of the students understood why the salt and ice get mixed. The next step was to draw and label a picture of the ice cream maker (see Figure 9.1). Students then guessed the purpose of these parts.

I gave students a list of the materials they would be using to build their own freezers: a bucket with a lid, ice, salt, newspaper, and masking tape. In their groups, the students had to design a freezer using these materials. Each group shared their plan and their thinking behind that design with the rest of the class. Groups were then allowed to revise their plan based on the shared ideas. Then came the fun part—building the freezer. Students tested the success of their freezer by simply freezing their juice pops. If the juice turned into ice pops, the freezer design was a success. If their freezer was not as successful, students had another opportunity to work with the group to decide how they could change their plan to improve their freezer.

I was able to evaluate many things when I looked over their designs, revisions, and conclusions in their science notebooks. I was able to determine the level of understanding regarding

FIGURE 9.1

Student Notebook Sample From Solids, Liquids, and Gases Lesson

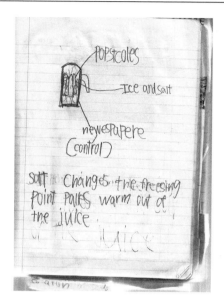

insulation, the purpose of air in the design, as well as the energy transfer needed to change liquid matter into solid. By altering the lesson in the ways mentioned above, I was also able to evaluate my students' ability to think about a problem critically, create a way to solve a problem, and test a prediction. The rubric used for this evaluation can be found online (see NSTA Connection, p. 86).

Sound

There aren't many opportunities in my program's engineering of sound exploration for students to design and build something to solve a problem. However, one of the science and engineering practices included in the *NGSS* states that students in second grade must "ask questions based on observations to find more information about the natural and/or designed world" (NGSS Lead States 2013). In the lesson about

pitch, the lab activity for my program gives me a list of steps through which to guide my students in discovering that the characteristics of given objects affect their pitch. Again, this lesson uses structured inquiry to tell the students what to do and how to do it and then expects students to ask questions later. There is no opportunity for the students to ask their own questions, investigate, and produce their own conclusions based on their observations.

Reflecting on this idea of student-generated questions and student inquiry, I decided to try the lab activity a little differently. Rather than guiding students through each step as listed in the lab manual, I simply gave them a challenge. I told them they were going to get a set of materials (metal tubes of different lengths, a wooden stick, and foam padding) that they needed to use to try to explain pitch. See NSTA Connection (p. 86) for a full lesson plan.

Students were put into groups and required to list some questions they had in their science notebooks regarding pitch and sound. Once the students generated those questions, they had the opportunity to discuss and share them with other students in the class. Some questions generated by my students include the following:

- What is pitch?

- Why are they different sizes?

- What kind of sounds will they make?

- Do you hit the tubes with the stick or with other tubes?

- What is the foam pad for?

- Does it matter how hard we hit the tubes?

They could revise their questions or add new ones throughout this part of the lesson. After each student had a list of questions to test, the groups then had to devise a plan regarding how to test their questions. Once they decided on a plan, the students had a chance to investigate their materials and test their ideas. As I walked around the classroom, I noticed that some of the groups placed the metal tubes right on the table top while others placed them on the foam pad. Some groups used the rhythm stick to hit the metal tubes while others hit the metal tubes on each other. Before I could begin to ask some guiding questions, the students noticed what other groups were doing. They watched neighboring tables use the rhythm sticks and the foam pads and recognized the difference in sound generated. So, they began to try those ideas. They noticed the difference in sound. Some of the tubes made higher sounds and some made lower sounds. I noticed that groups started to place their tubes in order of size after comparing the sounds they made. We shared observations and worked together as a class to define *pitch* as how high or low a sound was. Observations were recorded on a chart in their science notebooks.

To satisfy the challenge, students had to create their own questions, plan how to test those questions, and revise their ideas. The documentation in their science notebooks (Figure 9.2) was used to assess student understanding regarding the definition of pitch, the features that affect pitch, and the ability to ask questions and design an investigation meant to answer those questions. The rubric (see NSTA Connection, p. 86) was used to assess the work in their science notebooks.

The fact that student inquiry guided the investigation rather than a lab manual or a series of steps to follow made the lesson so much more motivating. The students guided each other and were given a chance to think for themselves using background knowledge or their own misconceptions about pitch. They were willing to share their ideas with each other and try to satisfy the challenge.

FIGURE 9.2

Student Notebook Sample From Engineering of Sound Lesson

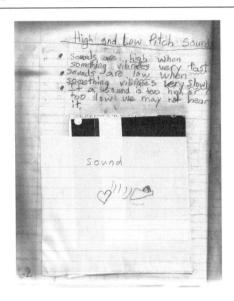

My Observations

As I walked around the classroom during this lesson and spoke to each group, I not only noticed 100% participation but also observed science-based conversation happening among the students. My second graders were challenging each other's ideas in the pursuit of understanding and the development of content knowledge. I heard questions such as, "Why do you think that?" and "How did you do that?" My students were explaining themselves to each other and using their investigations to demonstrate the use of the newspaper as insulation. When doing the ice cream maker lesson, one student told the rest of his group, "The newspaper is an insulator just like your coat." They were making connections between the science content and their world.

The ideas I have discussed are not much different from what I and most teachers have been doing in science classrooms for years. Highly engaging classrooms have been using lab activities to help model and illustrate concepts and content for a long time. But the introduction of the *NGSS* requires us to look at our lessons and make small but effective changes to incorporate these engineering principles. It is about finding ways to incorporate guided inquiry into the lesson format rather than relying on structured inquiry. The *NGSS* are about creating individuals that can face real-world problems and generate solutions to those problems to further society and technological advancement. The introduction to the *NGSS* states the following:

Implementing the *NGSS* will better prepare high school graduates for the rigors of college and careers. In turn, employers will be able to hire workers with strong science-based skills—not only in specific content areas, but also with skills such as critical thinking and inquiry-based problem solving. (NGSS Lead States 2013, p. xv)

I hope I have been able to illustrate the importance of beginning by making small but important changes to our lab activities in our science classes, even at the elementary level. With a slight shift in lesson design and focus, some extra materials, and a few extra minutes, we can help our students become critical thinkers and active problem solvers and move toward the performance expectations found in the *NGSS*.

Connecting to the *Next Generation Science Standards*: Solids, Liquids, and Gases

The materials, lessons, and activities outlined in this chapter are just one step toward reaching the performance expectations listed below. Additional supporting materials, lessons, and activities will be required.

2-PS1 Matter and Its Interactions *www.nextgenscience.org/2ps1-matter-interactions* **K-2-ETS1 Engineering Design** *www.nextgenscience.org./k-2ets1-engineering-design*	**Connections to Classroom Activity**
Performance Expectations	
K-2-ETS1-1: Ask questions, make observations, and gather information about a situation people want to change to define a simple problem that can be solved through the development of a new or improved object or tool	Observed a variety of ice cream makers to generate questions about their design Formulated an idea about the purpose of each component of then ice cream maker
K-2-ETS1-2: Develop a simple sketch, drawing, or physical model to illustrate how the shape of an object helps it function as needed to solve a given problem	Drew scientific pictures to show how the ice cream maker was designed, including labels and ideas about the purpose of design features Based on investigation of ice cream freezers, designed a homemade freezer intended to freeze juice into ice pops

2-PS1 Matter and Its Interactions www.nextgenscience.org/2ps1-matter-interactions **K-2-ETS1 Engineering Design** www.nextgenscience.org./k-2ets1-engineering-design	**Connections to Classroom Activity**
Performance Expectations	
K-2-ETS1-3: Analyze data from tests of two objects designed to solve the same problem to compare the strengths and weaknesses of how each performs	Tested their group's own homemade freezer and compared results with other groups Compared freezer designs to identify aspects of successful homemade freezers
Science and Engineering Practices	
Asking Questions and Defining Problems	Asked questions about design aspects of homemade ice cream makers
Developing and Using Models	Designed and created homemade freezers based on observations
Analyzing and Interpreting Data	Tested the freezers for success and compared freezer structures and functions to that of other groups
Disciplinary Core Ideas	
ETS1.A: Defining and Delimiting Engineering Problems • Asking questions, making observations, and gathering information are helpful in thinking about problems.	Identified the purpose of freezers and designed them to meet the objective using what they knew about homemade ice cream makers
ETS1.B: Developing Possible Solutions • Designs can be conveyed through sketches, drawings, or physical models. These representations are useful in communicating ideas for a problem's solutions to other people.	Drew designs in their notebooks, explained how aspects would work to freeze the juice, and developed a functioning physical model based on design sketches
ETS1.C: Optimizing the Design Solution • Because there is always more than one possible solution to a problem, it is useful to compare and test designs.	Redesigned freezer after observing and analyzing the success of others' freezers
2-PS1.B: Chemical Reactions • Heating or cooling a substance may cause changes that can be observed.	Created a solution and changed the substance from a liquid to a solid or semi-solid by cooling
Crosscutting Concept	
Structure and Function	Related the design features of the homemade ice cream makers to their functions through asking and answering questions about them Analyzed the success of the homemade freezers and identified aspects that made them successful or aspects needing redesign

Source: NGSS Lead States 2013.

Connecting to the *Next Generation Science Standards*: Engineering Sound

The materials, lessons, and activities outlined in this chapter are just one step toward reaching the performance expectations listed below. Additional supporting materials, lessons, and activities will be required.

K-2-ETS1 Engineering Design *www.nextgenscience.org/k-2ets1-engineering-design* **2-PS1 Matter and Its Interactions** *www.nextgenscience.org/2ps1-matter-interactions*	**Connections to Classroom Activity**
Performance Expectation	
1-PS4-1: Plan and conduct investigations to provide evidence that vibrating materials can make sound and that sound can make materials vibrate	Designed a method to test the questions generated about how the different metal tubes would produce different pitches Maintained observations from tests in student-generated tables or charts
Science and Engineering Practices	
Asking Questions and Defining Problems	Generated a list of questions about pitch and how different attributes of the metal tubes affect pitch
Planning and Carrying Out Investigations	Decided on a series of steps to conduct tests and answer the questions the class posed
Disciplinary Core Ideas	
ETS1.A: Defining and Delimiting Engineering Problems • Asking questions, making observations, and gathering information are helpful in thinking about problems.	Generated a list of questions about pitch and how different attributes of the metal tubes affect pitch Made observations about the pitch of the different-sized tubes and ranked them Compared results of pitch ranking and discussed irregularities between groups
2-PS1.A: Structure and Properties of Matter • Different properties are suited for different purposes.	Investigated the provided materials to determine how they affect the type of sound that can be produced
Crosscutting Concepts	
Cause and Effect	Used observations and ranking order to confirm or refute ideas about pitch, as affected by physical attributes
Structure and Function	Discovered through experimentation that the physical structure of the tubes affects the rate of vibration, which affects pitch

Source: NGSS Lead States 2013.

References

NGSS Lead States. 2013. *Next Generation Science Standards: For states, by states.* Washington, DC: National Academies Press. *www.nextgenscience.org/next-generation-science-standards.*

Zion, M., and R. Mendelovici. 2012. Moving from structured to open inquiry: Challenges and limits. *Science Education International* 23 (4): 383–399.

NSTA Connection

For the complete lesson plans and the rubric, visit *www.nsta.org/sc1503.*

Integrating Design

By Peggy Ashbrook and Sue Nellor

Teachers sometimes pose problems to engage children in engineering a solution, and sometimes children pose questions to teachers. Often, the problems children encounter in daily life are most important to them. They want to make their environment meet their needs. When wind blows their drawing paper away, they find a stone to weigh it down. If the sandbox or mud puddle is too dry for their purposes, they will try various ways to move water to it. Children may simply be enjoying the process, or they may be thinking of ways to improve it. By making a statement or a query, teachers foster the development of children's design process. Step in when frustration is not aiding their learning, and then focus their attention on an area of difficulty.

Engineering is such a common part of children's work in early childhood programs that teachers can simply look around the room to identify examples of students engaging in engineering practices. A list of a child's completed tasks might read as follows:

- Defined a problem: "My coat keeps falling down [from the hook]. The hook won't work."

- Developed a model: "It's not big enough. I need something bigger."

- Analyzed and interpreted data: "Lots of coats fall down."

- Used mathematics and computational thinking: "It needs to be much bigger."

- Engaged in argument from evidence: "No, it's not because I didn't do it right, the hook is too short for my thick coat."

- Designed a solution: "I'm going to tape this fat stick on my hook. It's longer so my coat will stay up."

Through trial and redesign, children discover if their solutions will solve their problems. This scenario would happen over time, along with discussions with other children and teachers. See NSTA Connection (p. 90) for additional examples of children using engineering practices. Teachers are cautioned to be aware of common misconceptions, such as children "building first" before understanding the problem and not noticing a problem with their design because they do not focus during the test of their design solution

10

Children can use similar materials indoors and outdoors to engineer design solutions to an everyday problem.

(Crismond et al. 2013). Children must have time to become familiar with the materials and the possibilities before they are ready to apply what they have learned to solving a problem. They may spend a long time exploring the properties of the materials before they begin making purposeful changes to refine their design so their design works best.

Integrate engineering design into your curriculum by building on the everyday problems that young children encounter in the school day. In the activity that follows, note the steps for redesign, and be open to following *a* design process, not a rigid *the* design process.

Moving Water

Objective

Engage children in designing a system to move water (or marbles) from one location to another.

1. To begin an engineering problem-solving activity, listen to children's conversations and discover problems that interest them. Children in this case were frustrated by the lack of a constant water supply in the Mud Box (see NSTA Connection, p. 90) because a bucket brigade was not bringing water fast enough to maintain the flow. Pose a question such as, "If it's not working, how can we get water from the source to the Mud Box without a bucket?" Children proposed using the available PVC pipes.

2. Try the design process by yourself to become familiar with problems that children might encounter. This prepares you to lead discussions to support children's thinking about the design process.

3. Have children identify the problem of moving water from the source to where it will be used. Support discussions by drawing possible solutions of designs using materials that are available to carry water. Indoors, children can model possible construction designs using PVC pipes and use marbles to represent water.

4. Provide the materials and describe how to carry the pipe sections safely (hold them vertically and close to your body) to prevent accidentally poking others. Introduce and name the pipes and fittings—90° elbows, tees, and wyes—at a sensory table with water as a center. Ask them to investigate how the different fittings affect the direction of the water flow.

5. In the following days and weeks, provide the materials outside for children to construct systems to enable water flow from the source to where they will use it. Children may have difficulty adjusting the end fittings or have arguments about which design will be most successful. Design problems may be related to the length of the pipe, direction, or the elevation of the structure. To support children's engineering design work, closely observe to notice problems and frustration with the process. Ask yourself when to insert yourself into the children's work and how to frame your comments or questions. Are the children still exploring the properties of the materials, or are they making purposeful changes to refine their design so their design works best? Have children wash their hands after handling muddy water.

6. Throughout the process, be aware that children continually incorporate new ideas or concepts into the whole plan. Outside, one child stated that to make the water travel past an incline in the pipes they had to "turn the water up. The more water there is, the bigger the hill you can have." Modeling the outdoor design indoors, using marbles and the same PVC pipe, helps children—even in a failed attempt—to understand how the change in direction of the pipes affects the outcome. They then incorporate this knowledge into their outdoor design.

Failed attempts are valuable learning moments. Give children opportunities to determine why their system does not work. They often misidentify reasons for failure and need to revisit their design and make adjustments to their structure. While working in a group, different skills and thinking processes support the group's efforts. Have materials available for exploration on a daily basis. Being familiar with the materials leads children to use them to solve other problems in the future.

Materials

- Water source or marbles
- PVC pipes, with 1 in. diameters in various lengths from 20 cm to 1 m long
- PVC fittings: 90° elbows, tees, and wyes (not threaded)
- Funnels for water
- Cups or buckets to catch the water or marbles
- Drawing materials

Reference

Crismond, D., L. Gellert, R. Cain, and S. Wright. 2013. Minding design missteps. *Science and Children* 51 (2): 80–85.

Internet Resource

K-2-ETS1 Engineering Design
www.nextgenscience.org/k-2ets1-engineering-design

NSTA Connection

Download a list of additional resources and other examples of engineering design in the classroom at *www.nsta.org/SC1509*.

11

Elephant Trunks and Dolphin Tails

An Elephant-Trunk Design Challenge Introduces Students to Engineering for Animals

By Lukas Hefty

"How do engineers help animals?" Mrs. Hefty, a second-grade teacher at Douglas Jamerson Elementary School in St. Petersburg, Florida, begins many lessons with this type of leading question. Talking about engineers and how they influence society is routine at Jamerson, a center for engineering and mathematics. All students in kindergarten through fifth grade engage in teacher-created, integrated engineering units of study, purposefully aligned to the *Next Generation Science Standards* (*NGSS;* NGSS Lead States 2013) for engineering design and the *Common Core State Standards* (*CCSS;* NGAC and CCSSO 2010) for English language arts and mathematics. The school's vision, "Engineering innovative thinkers for global success," comes to life as students develop the habits of mind—curiosity, creativity, critical thinking, perseverance, and communication—that successful innovators possess. The centerpiece of the curriculum is the Jamerson Engineering Design Process (see Figure 11.1), through which students collaborate to identify real-world problems, *plan* multiple solutions and select the most efficient one, *design* models and prototypes to *check* against rigorous design constraints, and *share* their findings.

FIGURE 11.1

Engineering Design Process

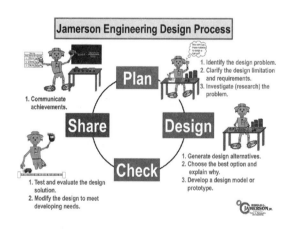

"How do engineers help animals?" Mrs. Hefty repeats. The students seem unsure and confused. In kindergarten and first grade, they were exposed to engineers who design bridges and skyscrapers, "just right" chairs, and aerodynamic ships. But animals? After some time to brainstorm together, they come up with a few ideas: engineers may help veterinarians heal sick animals or zookeepers develop safe habitats. Mrs. Hefty reads aloud the nonfiction text *Pierre the Penguin* by Jean Marzollo (2010). After molting, as all penguins do, Pierre's feathers did not grow back. Engineers at the California Academy of Sciences designed a wetsuit to keep Pierre warm, with an unexpected side effect—Pierre grew his feathers back. After reading, the class watches a brief news clip about Pierre and views the California Academy of Sciences live penguin webcam, learning more about penguin behaviors, such as preening and molting. The class uses the Jamerson Design Process to identify Pierre's main problem and the steps taken by engineers to develop a solution. The students are hooked, excited to discover more about how engineers help animals.

Mrs. Hefty opens the next day with a related question: "Has anyone heard of Winter the Dolphin?" Hands shoot up as students begin to retell the movie *Dolphin Tale* (2011). Rescued from a crab trap off a Florida coast and without a tail, Winter had little chance of survival. While veterinarians and scientists at the Clearwater Marine Aquarium fought to keep Winter alive, engineers spent six months designing a prosthetic tail, inventing brand-new materials and working through more than 100 prototypes. Winter's successful recovery has led to breakthroughs in prosthetic technology used to help returning war veterans. The class views a news clip outlining Winter's story and examines photographs of the prosthetic tail. In groups of three, the students use *www.seewinter.com* to investigate how engineers used each step of the engineering design process to help Winter. What stands out most for students are the number of models developed before they found one that fit just right. "Those engineers really had to persevere," one boy remarks. Perseverance, the understanding that *failure* is an opportunity to redesign, is a concept not lost on Jamerson teachers.

The Engineering Design Challenge

"Now it's your turn to be the engineer!" Mrs. Hefty announces excitedly. "We have a problem: The zoo just rescued an elephant that doesn't have a trunk!" The classroom buzzes with excitement. "Why would an elephant without a trunk be a problem?" The students share ideas about the importance of the elephant's trunk: smelling, grabbing objects, and so on. The class decides to research elephant trunk physical characteristics and functions at *www.seaworld.org*. The students are surprised at the number and variety of functions they discover: grasping, breathing, feeding, dusting, smelling, drinking, lifting, sound production and communication, defense and protection, and sensing. The trunk contains an estimated 100,000 muscles, providing the ability to expand, contract, and move in all directions (crosscutting concept: Cause and Effect). In other words, an elephant without a trunk is no small problem. This aligns with the *NGSS* Engineering Design performance expectation K-2-ETS1-1: "Ask questions, make observations, and gather information about a situation people want to change to define a simple problem that can be solved through the development of a new or

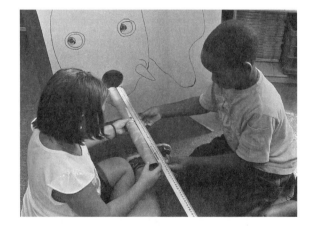

Two students measure the trunk.

improved object or tool," as well as disciplinary core idea ETS1.A: Defining and Delimiting Engineering Problems (NGSS Lead States 2013).

Mrs. Hefty continues, "Let's put on our engineer hats." She pauses for all of the students to "put on their hats." This week we are working as biomedical engineers, just like the ones who helped Pierre and Winter. Your challenge is to design a model prosthetic trunk for the rescued elephant, with the following design constraints:

- The trunk length must be 50–80 cm.

- The trunk must enable the elephant to breathe. (It should include "nostrils," or an opening that would allow for breathing.)

- The trunk must attach to the body (a science show board with an elephant body drawn and a hole for the trunk).

- Your team must use the Jamerson Design Process and the materials provided (science and engineering practice: Developing and Using Models).

Jamerson teachers consider ability levels, background and cultural experiences, and other special needs to intentionally form heterogeneous student teams. The teams examine the teacher-provided materials—paper towel rolls, construction paper, foil, string, and other craft supplies—and are also asked to brainstorm materials they could bring from home. One group decides to bring in a pool noodle to serve as the trunk. Using the design challenge handout (see NSTA Connection, p. 97) as a guide, the students begin talking about potential solutions. They review the physical characteristics and functions of an elephant trunk, consider the available materials, and draw individual sketches. Each child presents an idea, and the team discusses the positives and potential drawbacks. "Will that design

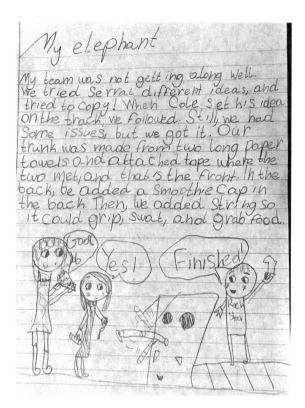

A student reflects on how well his team worked together.

allow the elephant to breathe? Could we use more flexible materials? How can we make it look more realistic?" (disciplinary core idea ETS1.A: Defining and Delimiting Engineering Problems; NGSS Lead States 2013) The formation of diverse student teams allows the students to support one another in developing the *best* idea. Eventually, each team comes to consensus and draws an agreed-on diagram. Over several days, the students work together to construct a model, check it against the design constraints, and make revisions to their diagram and physical model. They measure length and diameter and test for flexibility and strength. They are asked to consider factors such as whether the trunk will stretch and contract and the extent to which it will twist and bend without breaking (ETS1.B: Developing Possible Solutions; NGSS Lead States 2013).

Finally, they attach the trunk to the display board. Although each model looks different, every group reaches some level of success; engineering design performance expectation K-2-ETS1-2: "Develop a simple sketch, drawing, or physical model to illustrate how the shape of an object helps it function as needed to solve a given problem" (NGSS Lead States 2013).

A team attaches its trunk to the display board.

Like all good scientists and engineers, the teams share their findings, developing a written report and an oral class presentation. They consider points such as what did and did not go well, how the team cooperated, how many times they revised their model, and the final dimensions and materials used. The students are eager to not only share their displays but also see and hear what other teams discovered.

Science and Engineering

This series of lessons lasted approximately two weeks, 45 minutes per day, and was the conclusion to an integrated unit of study titled "Nature of Science and Engineering." Taught at the beginning of the school year, the unit focuses on the comparison of scientific and engineering practices in preparation for a year of similar,

real-world design investigations. It assists with building a collaborative, respectful classroom culture. The overarching goal of the K–5 integrated engineering curriculum is to develop in students the knowledge, skills, and habits of mind of successful scientists and engineers. While creating each unit of study, teachers consider the following:

1. *How can we level the playing field for students with varied background knowledge and learning styles?*

Beginning in kindergarten, Jamerson students are exposed to complex academic language. In many cases, vocabulary words are never explicitly taught but modeled by teachers during everyday engineering lessons. Over time, the students begin to appropriately use terms such as *prototype* and *design constraint*, discuss the metric system, and describe how they used each step in the engineering design process. In the vignette discussed, the use of articles, websites, and media exposed students to content in multiple ways. Techniques for online research and note taking were modeled, and responsibility was released gradually from the teacher to the students. Background knowledge was developed over multiple days before the design challenge, which leveled the playing field for students without prior experience with a topic, or students with a specific learning challenge. This increased collaboration and active participation from all students during the challenge.

2. *Does the engineering design challenge allow for creativity?*

Although all students use an engineering design process and are bound by the same design constraints, multiple solutions are possible and expected. Groups may achieve different levels of success, and two successful models may look very different from each other. Students with academic challenges or learning disabilities often do particularly well when given the opportunity to be creative.

3. *Does the design challenge encourage perseverance?*

Engineers in the real world will develop multiple, in many cases hundreds of, prototypes. They travel back and forth between design process stages, experiencing "failure" many times, which informs improvement and eventual success. The ability to persevere through setbacks and frustrations, design flaws, and disagreements with teammates develops character that transcends any single subject area. This may be the single most important benefit of exposing children to engineering design challenges in elementary school, and enough time must be allotted to allow for such repeated failure.

4. *How are collaboration and communication skills developed?*

Important to most jobs, those qualities are essential to an engineer's job. During the planning stage of the elephant trunk design challenge, the second-grade students began by thinking independently and then shared their ideas and sketches with teammates. Interestingly, the team didn't necessarily accept the "best" idea; rather, the team chose the idea that was most effectively communicated. Often the "gifted" learners came up with unique solutions but were not able to communicate them effectively to teammates—a skill that takes substantial practice to develop. Teams that had difficulty communicating were less likely to develop a design within the time constraint. During the share stage, students had to communicate results orally and in writing.

Over the next three years, these second graders will learn to publish their final reports, develop data tables and graphs, and create multimedia presentations to share their findings.

5. *How will student learning be assessed?*

Engineering design challenges at Jamerson are the culmination of a five-week unit of study, wherein students have the opportunity to apply learning to a performance-based task. The design challenge handout serves as a guide for students and includes a checklist or rubric based on the specific task's expectations. The scoring rubric (see NSTA Connection, p. 97) is shared with students when the challenge is introduced and includes categories such as teamwork, design product, and final report. For the elephant trunk design challenge, a separate section of the scoring rubric outlines the expectations for the written report.

When developing integrated engineering units of study, Jamerson teachers consider the following:

1. Which science core ideas will be the focus?

2. Which standards for engineering design will be included?

3. Which *CCSS English Language Arts, CCSS Mathematics,* and *CCSS Social Studies* standards can be naturally embedded?

4. How can we assess student learning (engineering design challenge)?

5. What knowledge and skills will the students need to be successful with the engineering design challenge (individual lessons)?

In beginning the seemingly daunting task of teaching engineering concepts to elementary-age students, the above steps can help, but the lead teachers at Jamerson also offer the following tips:

- *Start simple.* Begin with the science units you already teach, and add related engineering concepts and design challenges. This will ensure that learning is purposeful and standards-based rather than a set of disconnected engineering activities.

- *Collaborate.* Sharing the burden with colleagues will lessen the individual load and improve the product. It also provides a similar experience to what the students will face.

- *Reflect and revise.* Every engineering lesson will have unexpected outcomes, both good and bad. Like a design product, an engineering lesson will require several prototypes.

Conclusion

The Nature of Science and Engineering unit described sets the stage for a year of learning to collaborate and communicate with a diverse team using the engineering design process. It aligns with a coherent K–5 engineering curriculum that balances knowledge and practice. When we ask our kids to think like engineers—to define and design solutions to meaningful problems—we as teachers must make an uncomfortable shift by allowing our students to struggle and even fail. This requires a significant amount of time and patience, but those who make the shift over time will watch their students experience success through their failures and become highly motivated, thoughtful citizens.

Connecting to the *Next Generation Science Standards*

The materials, lessons, and activities outlined in this chapter are just one step toward reaching the performance expectations listed below. Additional supporting materials, lessons, and activities will be required.

2-PS1 Matter and Its Interactions *www.nextgenscience.org/2ps1-matter-interactions* **K-2-ETS1 Engineering Design** *www.nextgenscience.org/k-2ets1-engineering-design*	**Connections to Classroom Activity**
Performance Expectations	
2-PS1-2: Analyze data obtained from testing different materials to determine which materials have the properties that are best for an intended purpose	Tested different materials for strength and flexibility to determine which would be most appropriate for the model elephant trunk
K-2-ETS1-1: Ask questions, make observations, and gather information about a situation people want to change to define a simple problem that can be solved through the development of a new or improved object or tool	Gathered information about the characteristics and functions of elephant trunks to determine how best to construct a model
K-2-ETS1-2: Develop a simple sketch, drawing, or physical model to illustrate how the shape of an object helps it function as needed to solve a given problem	Developed a sketch and physical model to illustrate how the characteristics of an elephant's trunk help it perform many functions

2-PS1 Matter and Its Interactions www.nextgenscience.org/2ps1-matter-interactions **K-2-ETS1 Engineering Design** www.nextgenscience.org/k-2ets1-engineering-design	**Connections to Classroom Activity**
Science and Engineering Practices	
Developing and Using Models	Developed and tested a model elephant trunk against the design constraints
Constructing Explanations and Designing Solutions	Gathered information and designed an engineering solution for an elephant without a trunk
Disciplinary Core Ideas	
PS1.A: Structure and Properties of Matter • Different properties are suited to different purposes.	Tested and compared various materials in relation to their intended purpose as a model elephant trunk
ETS1.B: Developing Possible Solutions • Designs can be conveyed through sketches, drawings, or physical models. These representations are useful in communicating ideas for a problem's solutions to other people.	Developed diagrams and a model elephant trunk to meet the given design constraints
Crosscutting Concept	
Structure and Function	Explained how the shape and structure of an elephant's trunk affect its function

Source: NGSS Lead States 2013.

References

Dolphin tale. 2011. Film, directed by C. M. Smith. Burbank, CA: Warner Bros.

Marzollo, J. 2010. *Pierre the penguin.* Ann Arbor, MI: Sleeping Bear Press.

National Governors Association Center for Best Practices and Council of Chief State School Officers (NGAC and CCSSO). 2010. *Common core state standards.* Washington, DC: NGAC and CCSSO.

NGSS Lead States. 2013. *Next Generation Science Standards: For states, by states.* Washington, DC: National Academies Press. *www.nextgenscience.org/ next-generation-science-standards.*

Internet Resources

Clearwater Marine Aquarium
www.seewinter.com

SeaWorld
www.seaworld.org

NSTA Connection

Visit **www.nsta.org/SC1412** for the design challenge handout, scoring rubric, and a sample lesson plan.

12

Engineering Adaptations

Many of the Science Investigations We Use in Our Classrooms Already Can Be Adapted to Engage Students in Engineering Design

By Anne Gatling and Meredith Houle Vaughn

Engineering is not a subject that has histori-cally been taught in elementary schools, but with the emphasis on engineering in the *Next Generation Science Standards* (*NGSS*), curricula are being developed to explicitly teach engineering content and design. However, many of the scientific investigations we already con-duct with students have aspects of engineering design. This allows us as teachers to leverage not only existing materials but also our own expe-rience and expertise to incorporate engineering design into our classroom curricula.

By engineering design, we are referring to the process of students engaging systematically in an iterative process to solve real-world prob-lems. This process includes three components: defining and delimiting engineering problems, designing solutions, and optimizing the design solution.

In this chapter, we share two examples of "classic" science lessons we have adapted to teach science content through engineering design. We also provide some more general strat-egies for adapting science investigations to focus on engineering design.

Case 1: Moving Pigs

Inclines provide opportunities for young elemen-tary students to explore force and motion–based concepts at a very basic level. As part of inquiry-focused investigations, my preservice teachers would have their K–2 students in an after-school program build ramps that would be set up at various heights to release a ball or a car at fast or slow speeds. *Revealing the Work of Young Engi-neers in Early Childhood Education* (Van Meeteren and Zan 2010) inspired the idea for this lesson. The American Association for the Advancement of Science has a website called Science Net-Links, which provides excellent investigations complete with key questions to spur student thinking: Ramps 1: Let It Roll; Ramps 2: Ramp Builder; and Making Objects Move (see Internet Resources, p. 106). In addition to these resources, we would then also incorporate various materi-als to introduce the concept of friction, demon-strating how it influences the speed of the ball or small car on an incline. However, the *NGSS* performance expectation K-PS2-2 reads, "Ana-lyze data to determine if a design solution works

Inclines provide opportunities for young elementary students to explore force and motion–based concepts at a very basic level.

as intended to change the speed or direction of an object with a push or pull" (NGSS Lead States 2013, p. 4). This means teachers can extend this common activity into a more authentic learning experience by setting up an engineering design challenge such as an escape route for stranded pigs for students to design and build.

Define the Problem

Rather than just building a ramp from point A to point B, teachers can transform this common lesson into an engineering design challenge. In this particular example, the design challenge is as follows:

There are pigs (Ping-Pong balls or marbles) that have been trapped on a river island due to rising flood waters. Thankfully, there are still a few rocks poking above the water at various points between the pigs and the riverbank, but the rocks are too far for the animals to reach. How can you help the pigs get off the island?

Students can work together as partners to further identify the problem by asking questions about the situation to gather information so that they can effectively design a safe route for the pigs to escape. This lesson uses small- and large-group discussion, designing, testing, and rebuilding to start the cycle again, and it can generally be completed in an hour. Materials include cotton swabs, clay, cove molding (a wooden concave board that cradles a Ping-Pong ball perfectly and can be cut to varying lengths), blocks, and ping-pong balls or marbles. Final "bridges" can be built either on top of a group of student desks or on the floor. Blocks or other classroom items can serve as the "rocks" in the river. Student runners gather all materials except the cove molding, which teachers can distribute to the groups. Insturct students on the safety of using these materials and their own safety, specifying precautions they need to take around rivers or lakes, frozen or not. See Roy (2012) for information on using clay safely in the classroom.

The cove molding is relatively light but a bit awkward to move around. It is best for the teacher to distribute the cove molding to each group, placing it on the tables in the general alignment that the students will be using it. That way, students can easily just lift one end or the other to place blocks below.

Develop and Optimize Solutions

Once the scenario has been established, ask students, "What is the problem for these little pigs? How can we help them?" After they brainstorm with each other, ask, "What are your ideas? What are other groups' ideas? Let's share." The questions for this engineering design challenge were modeled after "Shedding Light on Engineering Design" (Capobianco, Nyquist, and Tyrie 2013). At this point, encourage students to share ideas of ways to solve the problem with each other and offer guidance on how to work to develop a plan for possible solutions. Next, distribute cotton swabs and clay and ask, "How can we build a model with these items to get the pigs safely across the water?" Students work together as a group to design a plan to get the pigs to safety by either drawing or building a segmented path with clay and cotton swabs and then sharing.

Once they have drawn or built their models, ask students to compare the models: "Which model seems to offer the best solution to getting the pig across the stream?" Sometimes there may be aspects of a few models that work together for the solution. Together, determine which model(s) would be best to build using the "real" materials—cove molding (available at most hardware stores) and blocks—to build a type of marble run. Depending on the quantity of your materials, you could have students build one or more models to test the most effective way to transport the pigs from the island to the riverbank. Have students work together to test their model and compare their models to those of other students, "How did your model work?" "Did your model work to get the pigs

from the island to the riverbank?" "Did something not work? Then, how will you improve your model? At this point, explain that engineers work together to find the best solution to a problem by developing models that they test and retest. Additional questions you could ask could include the following: Did you have to vary the speed at which the pigs had to go? Which solutions seemed to best solve the problem? Did more than one model work? What were the strengths and weaknesses of each? Students will then draw each of the models that worked, highlighting areas that made each model unique or different from the other models. They will then write a short description of the model they think carried the pigs most safely and why.

Assessment would be based on whether the pigs are able to be carried safely across the river, as well as on the students' ability to highlight the strengths and weaknesses of their own model and the other models. Last, share the type of engineer that would design a similar type of solution or model to solve a problem—in this case, a structural engineer. Modifying this force and motion lesson helped us retain the original intent of the content, force and motion, while incorporating the explicit focus of engineering.

Case 2: Point Pollution in a Watershed

In the upper elementary grades, a stream table or watershed model is used to teach students about topics such as soil erosion and pollution runoff in watersheds. In some instances, teachers might build (or purchase) these models for students to observe, or students might have the opportunity to observe these phenomena firsthand. For example, a lesson published on the Science Friday Education blog (Teachers Talking Science 2010) directs students to build two models out of sand; the first is a flat, inclined terrain, and the second is a model with hills and valleys to observe the formation of rivers. Another example from National Geographic instructs students to build a model out of clay, with the emphasis on helping students visualize point and non-point pollution (Walk n.d.). These activities have the explicit goal of teaching content and, possibly, an implicit goal around modeling. In working with preservice teachers and informal educators, we have modified this lesson to focus more explicitly on engineering design. Below we describe how we adapted a classic elementary lesson on watersheds, focusing on point pollution in watersheds and incorporating engineering design.

Although we could make several modifications to this lesson, we in particular identified modeling how point sources of pollution move into and through a watershed. By point pollution, we mean pollution originating at a single point, such as a sewage outfall pipe or factory. This is in contrast to non-point pollution, which is pollution entering a watershed from a more diffuse area, such as excess fertilizers or pesticides from agricultural or residential areas, sediment from construction sites, or oil and grease from roadways in urban areas.

Define Specific Criteria and Constraints

The challenge here is to modify the lesson from an explicit focus on modeling how point and non-point pollution move through a watershed to one that focuses on mediating pollution.

Initially, we followed the activity whereby students first observe the movement of point pollution in an "urbanized" watershed. The watershed model is similar to those described in the Science Friday Education blog (Teachers Talking Science 2010), using a mixture of rocks, soil, and sand to build a hill in a foil tray. The hill is then partially covered in foil to represent a "paved" area. Food coloring is introduced to the top of the hill as pollution, and a watering can serves to mimic rainfall into the watershed. Students quickly observe the pollution traveling down the paved surface and into the watershed. The challenge then is enlisting students to explore ways they can modify the watershed to contain sources of pollution.

Develop Solutions: Research and Explore

Students are then provided with resources to research possible solutions. The Environmental Protection Agency's (EPA; see Internet Resources, p. 106) website proved a useful first source of information on best practices for managing storm water. Students selected one or two strategies and explored how they worked and how they might be tested in our existing foil-tray model. Students examined available materials and how they could be used to mimic the strategies on the EPA website or used to inspire their own ideas to contain the storm water and capture the point source pollution. A variety of materials were available, such as sponges, straw, foil, and sand. For example, one group might opt to install permeable pavement (e.g., foil with small holes poked into it) laid over sand to serve as an initial filter for pollution. Students then presented their original sketches to their classmates for feedback and made modifications.

Optimize: Improve a Solution

After discussion and revision, students built their models according to their modified sketches. They then tested them, making notes about how pollution was contained and where their solutions failed.

Students shared these results with their classmates. They were then asked to reconsider their solutions and how they would revise their original sketches given what they learned during their tests. Students were assessed based on the effectiveness of their model to contain the pollution in the specified area, the quality of their design revisions, and the connections made to EPA-supported measures to address point source pollution. By modifying this lesson, we were able to both retain the original goal of teaching students about the movement of point pollution in a watershed and added to the explicit focus on engineering design.

Strategies for Modifying Lessons

Select an Appropriate Lesson

Select a lesson that can lend itself to a real-life problem. The lesson you select needs to be a problem that students can reasonably tackle. Here are two things to consider:

1. *Consider the degree to which students can connect and visualize the problem.*

In the first example, the idea of moving animals across a river is an idea young students can grasp. In the second example, we first needed to model the movement of point source pollution in a watershed before they engaged in the design activity. However, that modeling made the problem clear.

2. *The lesson needs to lend itself to being testable.*

Materials should be readily accessible so students have an opportunity to build and test their designs, ideally in multiple cycles of design and optimization.

Consider the Engineering Design Process

Engineering design incorporates the process of defining a problem, developing solutions, and optimizing the design solution. Consider how you will engage your students in grade-level appropriate activities (see NGSS Lead States 2013, Appendix I). As you incorporate these activities, it is critical to remember that these are components of engineering design, not a set of linear steps. In both examples, students engage in cycles of design and optimization.

Keep the Science Content

As we redesign lessons, one of the powerful aspects of engineering design is that it provides a real-world context for learning science. Although the adaptation of the lesson is foregrounding engineering design, don't forget about supporting students' conceptual understanding of the underlying science ideas as well. In the watershed example, students should walk away with an understanding of point pollution sources *in addition* to a richer understanding of engineering design!

There are many wonderful lessons available that could be adapted to incorporate engineering design. Consider selecting one of your favorite lessons to adapt this way. You will find that with time and experience it becomes easier to incorporate engineering design into lessons you already have.

Acknowledgments

Special thanks to preservice teachers Courtney McGowan and Kelsey O'Neil, who helped students design and build paths to carry the pigs safely across the river.

Connecting to the *Next Generation Science Standards*

The materials, lessons, and activities outlined in this chapter are just one step toward reaching the performance expectations listed below. Additional supporting materials, lessons, and activities will be required.

K–2 Forces and Interactions: Pushes and Pulls *www.nextgenscience.org/kfi-forces-interactions-pushes-pulls*	Connections to Classroom Activity
Performance Expectation	
PS2-2: Analyze data to determine if a design solution works as intended to change the speed or direction of an object with a push or a pull	Planned and designed a pathway and then conducted an investigation to push the pigs to safety
Science and Engineering Practices	
Planning and Carrying Out Investigations	Planned and constructed models of a path to carry pigs safely to the other side of the river
Analyzing and Interpreting Data	Recorded and shared observations of models that succeed in safely carrying the pigs Analyzed information to determine if there are ways the path works as intended
Disciplinary Core Ideas	
PS2.A: Forces and Motion • Pushes and pulls can have different strengths and directions. • Pushing or pulling on an object can change the speed or direction of its motion and can start or stop it.	Explored the varying amounts of strength they needed to push the pig so that it stays on the path and completes the full distance
ETS1.A: Defining and Delimiting Engineering Problems • A situation that people want to change or create can be approached as a problem to be solved through engineering	Planned, designed, and built their own model (as a group) to save the pig
Crosscutting Concept	
Cause and Effect	Determined the amount of force they need to use to keep the pigs on the path: too hard and the pigs fall off the path, too light and the pigs may not reach their destination

Source: NGSS Lead States 2013.

Connecting to the *Next Generation Science Standards*

The materials, lessons, and activities outlined in this chapter are just one step toward reaching the performance expectations listed below. Additional supporting materials, lessons, and activities will be required.

5-ESS2 Earth Systems *www.nextgenscience.org/5ess2-earth-systems*	**Connections to Classroom Activity**
Performance Expectation	
5-ESS2-1: Develop a model using an example to describe ways the geosphere, biosphere, hydrosphere, and/or atmosphere interact	Observed the interaction of water and land and researched and explored ways to modify the watershed to contain sources of pollution
Science and Engineering Practices	
Developing and Using Models	Planned and constructed models of storm water runoff
Constructing Explanations and Designing Solutions	Constructed explanations regarding the movement of pollution in the models and sought feedback from peers Observed characteristics of the pollution in the foil-tray models
Disciplinary Core Ideas	
ESS2.A: Earth Materials and Systems • These systems interact in multiple ways to affect Earth's surface materials and processes.	Evaluated the foil-tray watershed example
ETS1.B: Developing Possible Solutions • Research on a problem should be carried out before beginning to design a solution.	Researched EPA's best practices for managing storm water Designed strategies to contain point source pollution
Crosscutting Concept	
Systems and System Models	Used models to draw and explain how the pollution was contained and where their solutions failed

Source: NGSS Lead States 2013.

References

Capobianco, B., C. Nyquist, and N. Tyrie. 2013. Shedding light on engineering design: Scientific inquiry leads to an engineering challenge and both are illuminated. *Science and Children* 50 (3): 58–64.

NGSS Lead States. 2013. *Next Generation Science Standards: For states, by states.* Washington, DC: National Academies Press. *www.nextgenscience.org/ next-generation-science-standards.*

Roy, K. 2012. Safety first: Modeling safety in clay use. *Science and Children* (50) 4: 84–85.

Teachers Talking Science. 2010. Stream table. Science Friday Education. *www.sciencefriday.com/ educational-resources/stream-table.*

Van Meeteren, B., and B. Zan. 2010. Revealing the work of young engineers in early childhood education. Collected Papers from the SEED (STEM in Early Education and Development) Conference. *http://ecrp.uiuc.edu/beyond/seed/zan.html*.

Walk, F. H. n.d. In your watershed: How do people impact a community's watershed and its freshwater supplies? *National Geographic. http://education.nationalgeographic.com/education/activity/in-your-watershed/?ar_a=1.*

Internet Resources

AAAS Science Netlinks: Making objects move
http://sciencenetlinks.com/lessons/making-objects-move

AAAS Science Netlinks: Ramps 1: Let it roll
http://sciencenetlinks.com/lessons/ramps-1-let-it-roll

AAAS Science Netlinks: Ramps 2: Ramp builder
http://sciencenetlinks.com/lessons/ramps-2-ramp-builder

Environmental Protection Agency
www.epa.gov

NSTA Connection

Download the *Common Core State Standards* Connections for this lesson at www.nsta.org/SC1509.

13

Sailing Into the Digital Era

Students Transform STEM Journals Into E-books While Integrating Science, Technology, and Literacy

By Janet Bellavance and Amy Truchon

I've never thought of myself as tech savvy, which may be surprising once you read what my students were able to accomplish. Given that I have 31 years of teaching experience, I probably fit more in the category of "digital dinosaur!" For this project, I teamed with a young colleague who is a "digital native," but labels don't define us. As science educators, it is ultimately our judgment that matters. My years of classroom experience are grounded on a solid foundation of child development and understanding of the writing process and science content. I know I can keep improving my teaching. About a year ago, I started teaching with the curriculum Engineering is Elementary (EiE), which engages elementary students in engineering simple technologies (for example, a parachute or a simple solar oven). They design, test, and then, based on test results, improve their design using an age-appropriate five-step engineering design process (EDP; Figure 13.1). This simple process—Ask, Imagine, Plan, Create, and Improve—turned out to provide an effective structure for me as a teacher as I worked to redefine what it means for students to create science journals.

FIGURE 13.1

Engineering Design Process

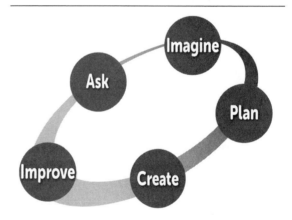

Ask

"What if our second-grade scientists could use digital technology (videos, voice recordings, photos) to review their test results and make thoughtful design improvements?" That's the question I asked myself while preparing to teach an EiE unit this past January. I had recently taken a professional development course with a team of colleagues on integrating iPads in the classroom, and each of us had received a "six pack" of iPads. I was determined to use this technology to transform learning in my classroom.

My students were already familiar with creating journals—collecting data, recording observations, and using this information to improve their design work—from their previous experience with EiE. This experience with journaling led us to the next step in the EDP: "What would it look like for a second grader to create an electronic STEM journal?" I thought electronic journals had the potential to do more than just substitute for written journals—they offered advantages that would redefine the educational experience of journaling.

Imagine

I was planning to teach an EiE unit that integrates engineering with science lessons about wind and weather. Students would use what they've learned to design sails for a model boat. I recruited our district's technology integration specialist, my coauthor Amy, and we sat down to imagine how iPads could add to the experience.

The iPad is a natural fit with science and engineering lessons. Children can record their observations not just by drawing or making notes but also by videotaping, taking photographs, or making a voice recording. They can easily refer to these observations as they review, discuss, and refine their designs, and they can share their learning with a wider audience (peers, parents, the community, or other students anywhere in the world). Meanwhile, when they review their videos and photos, students see themselves as scientists engaged in practice.

We imagined children using the tablets as digital STEM notebooks in the form of e-books. We pictured young learners circulating around the room with their devices, taking photos of themselves as they selected materials and built sails, and documenting themselves engaged in scientific and

engineering processes. We also imagined young engineers recording conversations as they discussed the materials they would choose for their sail designs and the rationale for their selection. We imagined them drawing and labeling their sail designs. We imagined students working in teams to videotape their sail trials, and then being able to watch them again and again to make thoughtful design improvements. This enhanced ability to review the trials and see exactly what was happening would boost student learning in science and engineering, whereas the collaborative conversations and explanatory writing would meet the second-grade *Common Core State Standards* for informational text writing.

As educators and learners, *asking* and *imagining* push us to redefine learning. However, it's ultimately the *planning* that ensures these visions become a reality in the classroom.

Plan

Creating an e-book on an iPad might seem like an ambitious assignment for second-graders. We chose the apps Book Creator and Explain Everything for our e-books because those apps are kid friendly, high-utility tools and have versatile features which, once taught, can be used repeatedly and adapted to new content and user needs. These features include an audio recording tool; drawing tools; editing tools; video editing, photo editing, and importing capabilities; the ability to publish books and share across platforms; and more.

We decided to use storyboards, a familiar planning tool in task analysis, to divide the process into discrete parts and create a structure for student work. It took several planning sessions to design our storyboard template in Book Creator (Figure 13.2). We used the steps of the engineering design process (Ask, Imagine, Plan, Create, Improve) as headings for each page in the digital journal. To further support student learning, we embedded prompts in the template, in the form of questions or directions as follows:

- Plan: Draw a diagram and label materials here.

- Create: Build and test your sail, put photo of sail and video of trial 1 here.

- Improve: What changes did you make, and why? Put video of trial 2 here.

FIGURE 13.2

An E-book Storyboard

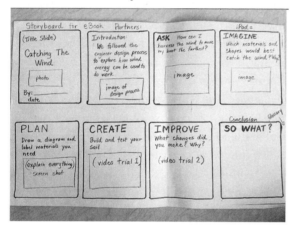

Finally, we incorporated required elements from the *Common Core State Standards* for informational text writing into the storyboard: introduction, facts and definitions, and conclusion.

We also wanted to make sure students were competent using the iPads, so Amy and I planned and taught lessons on how to take pictures, shoot video, and create folders to save the work. All students practiced videotaping and photographing before we started our engineering activity. All iPads were secured in padded, protective cases, and students were instructed on how to safely plug the devices in at recharging stations.

Book Creator has a number of different tools, and we did teach students how to use them.

However, we found that often we simply stood and watched as students explored the app and taught each other new things they discovered ("Oh, you can crop that photo. I'll show you how." Or "If you press this, you can move your text or bold a word." Or "*Trial* is an important science word, you should put that in a glossary.").

One other benefit of this project was the way it introduced students to responsible "digital citizenship." We consulted the National Educational Technology Standards and taught lessons on digital citizenship and how to ethically use online information. Students learned to ask permission before taking someone's photo, and that it's not okay to copy any image you find on the internet (instead they chose copyright-friendly images, using the app Haiku Deck). The students also learned about giving credit for the ideas of others, specifically with the EiE EDP graphic, which was an important visual that all the students included in their e-books.

Create

Now we were ready to release the responsibility to the children. When the students were ready to start designing their sails, they worked in teams of two. We assigned the partnerships based on a number of factors, including complementary abilities and social skills.

Students considered and selected different materials for their sails, including felt, foil, tissue paper, waxed paper, and card stock—all of which are inexpensive and easy to source. In addition to choosing materials, they could also make decisions about the sail's size and shape. For boat hulls, we used the kind of Styrofoam trays you get with some supermarket fruits and vegetables. For masts, we used wooden craft sticks.

After designing the sail and installing it on their boat, the students tested it under controlled conditions to see how far it would sail. We didn't need a pond or river to sail these boats! We made a track from two pieces of heavy gauge fishing line. Each boat attached to the track with soda-straw runners (the fishing line could be threaded through each runner before it was pulled taut between two classroom desks; see the straw placement in Figure 13.3). A household fan provided the wind. We found that a well-designed sail can move a boat dramatically fast and far. Eye protection is required for all stages of this activity (setup, hands-on activity, and take down). Make sure the fan is anchored or weighted down to prevent falling over. Remind students to never stick anything (e.g., fingers, pencils, or pens) into the moving blade and to keep loose objects away from the front of the fan.

FIGURE 13.3

One Student's Sail

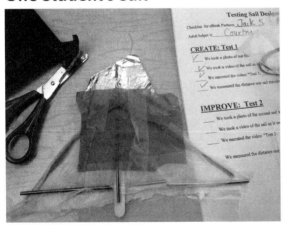

There was a lot going on in the classroom the day of the trials, so we recruited some parent volunteers to help us with tasks such as setting the boats on the tracks and turning the fan on and off. To help students collect their data (distance traveled) efficiently, we pre-marked the track with red lines at one-foot intervals; students used tape measures to measure the inches.

Students photographed their sails to record their shape and then took turns videotaping the trials, labeling each segment of videotape with their names and the trial number ("Trial 2 by Hannah and Jacques") for easy identification during their e-book creation. They also collected measurement data that would drive their design improvements.

We set—and enforced—a strict hands-off policy for adults with regard to digital technologies; the children did all the photographing and videotaping themselves. This was just one more example of applying the EDP. If the photos or videos didn't turn out that great, students could shoot more—and improve their technology skills in the process.

When the partners sat down to evaluate the test results, it was easy for them to replay their videos as much as they wanted to review their sail's performance. This was the main educational advantage of digital note booking—students could replay their test results (e.g., the boat tipped over, didn't move, moved slowly, or zipped down the track) carefully and repeatedly to see what factors influenced how far their sail traveled. They could connect the distance the sail traveled (data) with their design choices (material, size, shape).

Partners made improvements to their design—changing the material or the shape of the sail—based on their findings, then repeated the trial. Throughout the process, the students documented everything digitally. They labeled the components of a sail using the Explain Everything app, made predictions about sail performance, recorded observations and data from the testing process, and reported their conclusions as a summative assessment of their learning.

Students built their collaboration skills both in designing their sails and in creating their e-books using photos, videos, and storyboards (Figure 13.4). They negotiated taking turns working on pages, typing, recording, and working on layout decisions (i.e., placing text,

FIGURE 13.4

Sample E-book Pages

 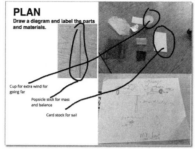 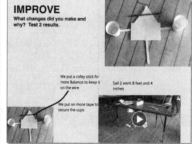

creating captions for photos, adding voice, using the "draw" tool to label or clarify, and justifying color choices). Electronic editing freed students from the burden of erasing smudgy pencil, and they could ask for help with typing if needed.

Students used a rubric based on the Common Core Standard for informational text to evaluate their eBooks (Table 13.1, p. 112; see NSTA Connection, p. 114). Using two mirroring software packages, AirPlay and Reflector, students could easily share content from their iPads on a large screen in our classroom, and we encouraged this sharing so that the teams could learn from one another.

This climate of organic sharing and "kids learning from kids" pervaded the classroom, with the teacher as a true facilitator checking in with the student teams, which were all moving at their own pace. I feel the high level of student engagement and creativity was a direct result of students taking ownership of the process, understanding the content and technology, having the ability to move at their own pace, and capturing themselves in the act of science. All students (English language learners, individualized education program students, behaviorally challenged students, gifted learners, and so on) were able to participate successfully in this science project.

Book Creator makes it easy for students to share their work in the book-sharing app iBooks or across platforms using the export as a video feature. So after we completed the unit, parents and community members were invited to an "eLearning Adventures" celebration where each set of partners shared their STEM journals. Our eBooks are now available for anyone to review (see Internet Resource, p. 114)!

Improve

Where do we go from here? Our next unit of study is called personal interest projects (PIPs). Children choose a subject they are interested in learning more about, ask an inquiry question, do research, and choose a method to present their learning to their peers. Some children are creating traditional posters and models, but it is exciting to see how many students are choosing to use e-books to, for example, teach others about how diamonds are formed or create an electronic quiz about the Tasmanian devil. It's been a natural progression from hand-written STEM journals to students directing their own learning and using e-books to share their learning. In sum, digital journals are here to stay. Our classroom is in a state of transformation where learning is being redefined, thanks to a new technology, and the willingness to ask, "What if … ?"

TABLE 13.1

E-book Rubric

Score 1 Getting Started	Score 2 Almost There	Score 3 Got It!	Score 4 Wow!
My book has a cover.	My book has a cover with a title, authors, and photo.	My book has a cover with a title, authors, and a photo that shows my topic.	My book has a cover with a title, authors, and a photo that shows my topic.
My book has photos.	My book has some photos, but they don't go with my topic. The photos aren't clear.	My book has at least four clear photos that help the reader understand my content.	My book has more than five clear photos that help the reader understand my content.
My book does not have a video.	My book has a video. The video is unclear.	My book has one clear video that you can watch to learn more about my subject.	My book has one clear video with voice added to learn more about my subject.
My book has some information about my topic, but it does not answer my question.	My book has some information about my topic and answers my question.	My book introduces a topic, uses facts and definitions to develop points, and provides a concluding statement.	My book introduces a topic, uses multiple facts and definitions to develop points, and provides a concluding statement.
There are no text features in my book.	My book has 1–2 text features to help the reader understand my subject.	My book has 3–4 text features to help the reader understand my subject. (headings, diagram/labels, bold type, comparisons, cutaways, glossary, table of contents, maps, graphs)	My book has 3–4 text features to help the reader understand my subject. My book has a glossary.

Connecting to the *Next Generation Science Standards*

The materials, lessons, and activities outlined in this chapter are just one step toward reaching the performance expectations listed below. Additional supporting materials, lessons, and activities will be required.

K-2-ETS1 Engineering Design *www.nextgenscience.org/k-2ets1-engineering-design* **2-PS1 Matter and Its Interactions** *www.nextgenscience.org/2ps1-matter-interactions*	**Connections to Classroom Activity**
Performance Expectations	
K-2-ETS1-2: Develop a simple sketch, drawing, or physical model to illustrate how the shape of an object helps it function as needed to solve a given problem	Designed a sail and tested to see how far it moves a boat
K-2-ETS1-3: Analyze data from tests of two objects designed to solve the same problem to compare the strengths and weaknesses of how each performs	Used the test results to improve on sail designs
2-PS1-2: Analyze data obtained from testing different materials to determine which materials have the properties that are best suited for an intended purpose	Evaluated different materials that could be used to make a sail
Science and Engineering Practices	
Planning and Carrying Out Investigations	Worked collaboratively as a team to fulfill the design challenge
Analyzing and Interpreting Data	Tested different materials to understand their properties; chose materials and designed a sail; and based on test data, re-designed the sail and retested
Constructing Explanations and Designing Solutions	Made observations and provided explanations concerning design choices; communicated the choices and solutions in written and video format
Disciplinary Core Ideas	
PS1.A: Structure and Properties of Matter • Matter can be described and classified by its observable properties. • Different properties are suited to different purposes.	Examined, evaluated, and described different kinds of materials Used materials to build a sail, testing to see what materials are most effective for catching the wind
ETS1.B: Developing Possible Solutions • Designs can be conveyed through sketches, drawings, or physical models. These representations are useful in communicating ideas for a problem's solutions to other people.	Communicated the design process and solutions in written and video format using an iPad as the communication device
Crosscutting Concept	
Cause and Effect	Selected materials based on the effect they have on the functioning of a sail and used selections to create a boat sail shape that will most efficiently catch the wind; through analyzing data, made adjustments to the design using evidence gathered to support ideas

Source: NGSS Lead States 2013.

Reference

NGSS Lead States. 2013. *Next Generation Science Standards: For states, by states.* Washington, DC: National Academies Press. *www.nextgenscience.org/next-generation-sciencestandards.*

Internet Resource

Wind Ebooks Video Format Bellavance
http://bit.ly/1J7bW9r

NSTA Connection

Download the rubric at *www.nsta.org/SC1511.*

Inventing Mystery Machines

Collaborating to Improve Teacher STEM Preparation

By Shelly Counsell, Felicia Peat, Rachel Vaughan, and Tiffany Johnson

The call for STEM education warrants the need for teacher education programs to prepare highly qualified educators. Informal education is an important part of the solution. Collaboration with local museums can maximize learning experiences and outcomes for both inservice and preservice teachers with real-world applications (and lead to improved STEM instruction and learning for children). Based on the work completed between the University of Memphis's early childhood education (ECE) program and the Children's Museum of Memphis (CMOM), our insights can be used to improve classroom teachers' practices in the primary grades. This collaboration provided preservice teachers (concurrently enrolled in an early childhood math and science methods course) the unique opportunity to develop science units that could be implemented with young children at the local museum and in practicum placements in assigned classrooms. This arrangement increases preservice teachers' STEM knowledge, skills, and teaching practices by maximizing their opportunities for implementation beyond the immediate practicum experience. In addition,

collaborations between universities, museums, and local schools have the potential to build and reinforce important STEM partnerships that further bridge and coordinate STEM teaching efforts used to improve STEM outcomes for young children.

One STEM unit of study in physics, "Wheels in Motion: Mystery Machines," was developed by a team of four preservice teachers specifically to complement the CMOM's outreach program, Motion Commotion, which is a fast-paced program that challenges children to work in teams to design a gravity-powered car that travels the farthest distance. Wheels in Motion activities expand on physical science motion concepts that Motion Commotion addresses (now offered as part of the Motion Commotion outreach module for classroom teachers).

The Wheels in Motion activities encourage young children to explore and investigate the motion of wheels to strengthen and expand their foundational understanding of motion before creating gravity-powered cars. Three key approaches (and related practices) fundamental to early STEM learning with young children are fully used to improve teaching practices and learning outcomes:

- Learning Cycle Approach

- Inquiry Approach (Productive Questioning)

- Engineering Design Approach

Learning Cycle Pedagogy and Inquiry Practice

Each participating preservice teacher contributed one lesson that was organized and integrated into a comprehensive unit using adapted learning cycle pedagogy with three distinct phases—exploration, concept development, and application (Cooney, Escalada, and Unruh 2005)—which emphasize active, child-directed investigation. These lesson plans use familiar items and materials that can easily be implemented by classroom teachers.

Six types of productive questions (Elstgeest 2001) are used to support children's reasoning. Productive questioning provides a practical framework for teachers to help improve children's scientific thinking and active engagement. The need to acknowledge and respect children's thinking and ideas is equally paramount. Only through careful observation can teachers determine when to interject possible questions to facilitate learning without overwhelming the child's thoughts or interfering with his or her agenda (Counsell, Uhlenberg, and Zan 2013).

Engineering Design Process

Applying scientific thinking to create or invent something with an intended purpose or function (Cunningham 2009) uses a basic engineering design process (e.g., Ask, Imagine, Plan, Create, and Improve) that teachers can simultaneously

emphasize to further enhance science learning. This approach is likewise compatible with *A Framework for K–12 Science Education* (NRC 2012), which recommends eight practices of science and engineering. The Wheels in Motion unit includes multiple opportunities for children to plan and carry out investigations and analyze outcomes, as recommended by the *Next Generation Science Standards (NGSS)*. Opportunities to create something new with a designed purpose or function tap into children's critical thinking (the highest level of creativity on Bloom's Taxonomy, revised). Additional examples of productive questions and elaboration (as described later) help to highlight for teachers how they can use learning cycle phases, inquiry process, and engineering process design with young children. All lessons include an introduction and closure (in a condensed version for the purpose of this chapter).

Lesson 1: Wheels and Motion Introduction and Exploration Phase

Teachers set out large containers filled with different geometric shapes and solids (triangles, circles, squares, spheres, cubes, cones, and cylinders) for children to examine and explore. The teacher asks productive questions as children examine the shapes to help children compare and contrast each geometric solid's different mathematical attributes related to the kindergarten *Common Core State Standards (CCSS), Mathematics*, such as the size, number of angles, points, shape, and number of faces (e.g., a cube has six square faces) (NGAC and CCSSO 2010). Teachers guide and facilitate children's exploration and logical-mathematical thinking using productive questioning (What do you notice about the end of each cylinder? How many faces can you count on a cylinder? How are cylinders and spheres the same or different?). *CCSS Mathematics* and *CCSS English Language Arts* addressed during this investigation pertain to counting and cardinality, measurement, and geometry, as well as developing children's ability to ask and answer questions (see *CCSS* for kindergarten).

Next, teachers guide children through discussions and eventual investigations to explore different types of familiar wheels, such as a tricycle wheel, hamster wheel, roller skates, toy car, and toy tractor, as well as a hamster ball. Remind students to watch where they are walking to avoid slip or trip hazards if geometric shapes and solids happen to fall on the floor. Remind students to handle wheels on toys carefully, given they can have sharp edges and have oil on them. Students should wash their hands with soap and water after completing the activity.

Productive questions are used to guide and facilitate children's logical-mathematical and scientific thinking as they investigate relationships between different shapes and motions (see related *CCSS Mathematics* and *NGSS*).

- *Attention-focusing questions* (e.g., "What do you notice about the different wheels?"). Children's possible responses include, "The toy car and toy tractor wheels are black and made with rubber. The hamster wheel does not have rubber."

- *Measuring and counting questions* ("How many wheels are on a tricycle?"). Children respond, "A tricycle has three wheels. One big wheel and two small wheels."

- *Comparison questions* ("How are the hamster wheel and hamster ball the same or different?"). Children respond, "A hamster wheel and hamster ball are both round (circular). A hamster ball is the shape of a sphere. A hamster wheel is shaped like a flat cylinder."

- *Action questions* ("Can you make hamster wheels rotate around in all directions like hamster balls?"). Children respond, "Both hamster wheels and hamster balls roll when pushed. A hamster ball rolls in all directions. A hamster wheel will not roll when placed flat on the ground. A hamster wheel will only roll along its rounded edges."

- *Problem-posing questions* ("What happens if we take away the tricycle's front wheel?"). Children respond, "A tricycle with a missing front wheel is broken. The pedals won't work. The tricycle is stuck."

- *Reasoning questions* ("What would happen if tricycles used balls instead of wheels?"). Children respond, "A ball rolls in all directions. A tricycle with balls would be harder to direct (steer) where to go."

Lesson 2: Wheels and Motion Concept Development Phase

Teachers can continue guiding children's investigation of motion and related concepts (specifically, force, motion, and friction) using office chairs. Children are introduced to two types of chairs: an office chair with wheels on a pedestal leg and a second office chair with four legs and no wheels. Children examine the two chairs, noting the similarities and differences ("What do you notice about the office chair's pedestal legs?" "How many wheels does the office chair with a pedestal leg have?"). Children investigate what happens (subsequent motion) when they push and pull each chair on a flat (level) surface.

Children are able to explore the relationship between motion and surface by changing the flat surface from a tile floor to a carpeted area. Outdoors, the teacher helps children identify two places with an incline: a grassy hill and a cement ramp. Before chairs are released on grassy and cement inclines, it is important to discuss safety issues with young children. Discuss safety rules related to playground slides (e.g., waiting for children to get off the slide before the next child slides down) and relate those rules to this investigation ("Where do we want to stand when chairs are released on inclines?"). To ensure safety with young children, the teacher demonstrates this by sitting in the chair while another adult moves it. Pushing and pulling of each chair type on flat and inclined surfaces by students must be under direct adult supervision. Remind students to move chairs slowly one at a time to help prevent trip or slip fall hazards. Remind students to make sure other students are clear of the movement and direction of the chairs on flat and inclined surfaces.

Based on the children's prior experiences, children should make predictions about what will happen when each chair is released at the top of a small grassy incline (hill) versus cement incline (entrance ramp).

- *Attention-focusing questions* (for example, "What happens when the chair without wheels is released on the cement ramp?"). Children respond, "It skids. It falls over."

- *Measuring and counting questions* ("How far will the office chair with wheels roll?"). Children measure the distance by counting how many feet, from heel to toe (informal measurement). Children can also measure the distance using a piece of string (informal measurement) or a metric tape measure (formal measurement).

- *Comparison questions* ("What happened when the chair with four legs was placed at the top of each incline?"). Children respond, "The chair skidded before falling

over on the cement. The chair did not slide standing up on the grassy hill. It fell over at the top."

- *Action questions* ("Did the pedestal office chair roll differently when it was released on the grassy incline in comparison to the concrete incline?"). Children respond, "The office chair bounced as it rolled on the bumpy grass. The chair rolled faster on the concrete hill (inclined ramp)."

- *Problem-posing questions* ("How could we make the chair with four legs slide down the inclines?"). Children respond, "You could put a wheel on each leg. You could put the chair on a flattened cardboard box and slide it down the hill. You could try pushing it all the way down to the bottom."

- *Reasoning questions* ("What rule can we make about rolling chairs on different indoor floor surfaces?"). Children respond, "Chairs roll easier and faster on hard and smooth floors (surfaces). Carpet and rugs make it harder for wheels to roll. You have to push harder to make the wheels move."

Lesson 3: Varying Motion Application Phase

To expand on their understanding of inclines and ramps, teachers provide children with multiple sections of 1, 2, 3, and 4 ft. lengths of wood cove molding (1 ¾ in. wide, flattened bottom design purchased at any home improvement store) in the classroom block center, as recommended for use in the physical science module, *Ramps and Pathways* (DeVries and Sales 2011). Children use blocks (wood, foam, and cardboard), boxes, furniture, and other available items to prop up the cove molding sections to create ramp trajectories and inclines. Children release a variety of objects (such as matchbox cars, marbles, crayons, spools, Ping Pong balls, cubes, jingle bells, and pennies) to observe what happens (motion) as each follows the constructed ramp structures. Eye protection (sanitized safety glasses or goggles) is required for the setting up, hands-on exploration, and taking down phases of this activity. Remind students not to place marbles, pennies, and other small items in their mouths. Remind students to be careful not to walk on marbles. Have students wash their hands with soap and water after completing this activity.

As children investigate motion, continue employing productive questioning ("How does the small spool's motion compare with the marble's motion when each are released on your incline? What happens to the matchbox car when you raise your ramp incline higher? Does it matter whether objects roll or slide?"). Changing the elevation of the ramp allows children the opportunity to once again test their ideas.

Lesson 4: Mystery Machine Application Phase

Teachers and children gather an assortment of wheels (cardboard, plastic, rubber, wooden, and so on), miscellaneous items (washers, small cardboard boxes and cartons, wooden craft sticks, pipe cleaners, string or yarn, cotton balls, and rubber bands), and glue (non-toxic, low VOC). Eye protection (sanitized safety glasses or goggles) is required for the setting up, hands-on exploration, and taking down phases of this activity. Sticks and pipe cleaners have sharp ends. Remind students to handle them carefully so as not to scratch or puncture skin. Remind students not to place washers, cotton balls, and other small items in their mouths. Students should wash their hands with soap and water after completing this activity.

Challenge children to create their own mystery machine with wheels. The same productive questioning is used in this final learning cycle phase as children apply scientific thinking and engineering design to create mystery machines.

- *Attention-focusing questions* ("What happens if you reposition the wheel?")

- *Measuring and counting questions* ("How far apart do you need to position the wheels?")

- *Comparison questions* ("How well do the wooden wheels move on tile floor compared to the rubber wheels?")

- *Action questions* ("Will rubber wheels move faster or slower on carpeted surfaces?")

- *Problem-posing questions* ("Can you make a mystery machine that rolls using only one wheel?")

- *Reasoning questions* ("Which type of wheels do you think will move most efficiently on different surface areas?")

Throughout the unit, children can digitally photograph, draw, or write about the different investigations exploring wheels and motion in science journals. Venn diagrams are used to compare and contrast items and motions observed. KWL charts and graphs can help organize and summarize science concepts and relationships based on motion and shape. These student products serve as important evidence of ongoing learning, understanding, and performance included in individual student science portfolios. A formative assessment probe is used to evaluate children's understanding of the relationship between an object's shape and its subsequent motion (see Figure 14.1, p. 120).

The "Wheels and Motion: Mystery Machines" workshop illustrates how two entities (a

FIGURE 14.1

Formative Assessment Probe

Source: Keeley, Eberle, and Farrin 2005.

How Will It Move?

The list below involves situations that cause objects to move differently. The objects are *italicized*. Put an X next to the situations in which the *italicized* object will *undergo* a *rolling* motion.

___ **A** Pushing an *office chair with wheels* on a tile floor.

___ **B** Pushing an *empty plastic saucer* (sled) on its bottom side down a grassy hill.

___ **C** Letting go of a *rubber kickball* at the top of a slide.

___ **D** Hitting a *plastic disc* on an air hockey table.

___ **E** Placing a hamster in a *hamster ball* on a carpeted floor.

___ **F** Releasing a *wooden cube* on a wooden ramp in the block center.

___ **G** Gently kicking a *can of soup* on its side (horizontally) on the cement sidewalk.

___ **H** Flicking a *marble* with your thumb across a wooden tabletop.

___ **I** Placing a *penny* flat on its face-side down at the top of a propped-up cardboard (paper towel) tube.

Explain your thinking. Describe the "rule" or reasoning you used to decide whether the objects roll.

university and a local museum) can combine their program strengths and resources in ways that create an infrastructure of collaborative practice. This infrastructure is needed to help increase STEM outcomes for ECE teachers in ways that will ultimately maximize science-learning experiences for young children. Teachers should contact local children's museums and university teacher education programs to identify available services and resources that support 21st-century high-quality STEM experiences.

Connecting to the *Next Generation Science Standards*

The materials, lessons, and activities outlined in this chapter are just one step toward reaching the performance expectations listed below. Additional supporting materials, lessons, and activities will be required.

K-PS2 Motion and Stability: Forces and Interactions www.nextgenscience.org/kps2-motion-stability-forces-interactions **K-2-ETS1 Engineering Design** www.nextgenscience.org/k-2ets1-engineering-design	**Connections to Classroom Activity**
Performance Expectations	
K-PS2-1: Plan and conduct an investigation to compare the effects of different strengths or different directions of pushes and pulls on the motion of an object	Investigated how round objects and familiar objects (office chairs) with and without wheels moved on different inclines and flat surfaces
K-PS2-2: Analyze data to determine if a design solution works as intended to change the speed or direction of an object with a push or a pull	Measured the distance traveled by different office chairs on different surfaces and used the analyzed data to design a new object ("mystery machine") that changed speed or direction with a push or a pull on different surfaces
K-2-ETS1-2: Develop a simple sketch, drawing, or physical model to illustrate how the shape of an object helps it function as needed to solve a given problem	Constructed mystery machine models that demonstrate how the shape of an object helps it move with a push or pull on different surfaces
Science and Engineering Practices	
Developing and Using Models	Developed mystery machine models; used the models to compare how they move with a push or a pull on different surfaces
Constructing Explanations and Designing Solutions	Developed claims with evidence about how each mystery machine would moved, which mystery machine would travel the greatest distance with a push or a pull on different surfaces, and what evidence would support these claims
Planning and Carrying Out Investigations	Planned and conducted investigations to collect data about the distance different mystery machines traveled on different surfaces
Disciplinary Core Ideas	
PS3.C: Relationship Between Energy and Forces • A bigger push or a pull makes things speed up or slow down more quickly.	Had multiple opportunities to observe, explore, and investigate how different familiar objects (office chairs) with and without wheels, as well as newly created objects (mystery machines) moved with a bigger push or pull on different surfaces

K-PS2 Motion and Stability: Forces and Interactions www.nextgenscience.org/kps2-motion-stability-forces-interactions **K-2-ETS1 Engineering Design** www.nextgenscience.org/k-2ets1-engineering-design	**Connections to Classroom Activity**
Disciplinary Core Ideas	
ETS1.A: Defining Engineering Problems • A situation that people want to change or create can be approached as a problem to be solved through engineering.	Explored, investigated, and varied different materials and shapes to create the mystery machine, observe how it moved with a push or a pull on different surfaces, and make modifications as needed
Crosscutting Concept	
Cause and Effect	Compared a variety of shapes, materials, and motions of wheels and balls to develop an understanding of how they affect motion Used different surfaces to cause wheels to move faster or slower Raised or lowered ramp inclines to cause marbles to speed up or slow down

Source: NGSS Lead States 2013.

References

Cooney, T. M., L. T. Escalada, and R. D. Unruh. 2005. *PRISMS (Physics Resources and Instructional Strategies for Motivating Students) PLUS*. Fairfield, OH: Centre Pointe Learning.

Counsell, S., J. Uhlenberg, and B. Zan. 2013. Ramps and pathways early physical science program: Preparing educators as science mentors. In *Exemplary science: Best practices in professional development*, ed. S. Koba and B. Wojnowski, 143–156. Arlington, VA: NSTA Press.

Cunningham, C. M. 2009. Engineering is elementary. *The Bridge* 30 (3): 11–17.

DeVries, R., and C. Sales. 2011. *Ramps and pathways: A constructivist approach to physics with young children*. Washington, DC: National Association for the Education of Young Children.

Elstgeest, J. 2001. The right question at the right time. In *Primary science: Taking the plunge*, ed. W. Harlen, 37–46. Portsmouth, NH: Heinemann.

Keeley, P., F. Eberle, and L. Farrin. 2005. *Uncovering student ideas in science, volume 1: 25 formative assessment probes*. Arlington, VA: NSTA Press.

National Governors Association Center for Best Practices and Council of Chief State School Officers (NGAC and CCSSO). 2010. *Common core state standards*. Washington, DC: NGAC and CCSSO.

National Research Council (NRC). 2012. *A framework for K–12 science education: Practices, crosscutting concepts, and core ideas*. Washington, DC: National Academies Press.

NGSS Lead States. 2013. *Next Generation Science Standards: For states, by states*. Washington, DC: National Academies Press. *www.nextgenscience.org/next-generation-science-standards*.

Am I *Really* Teaching Engineering to Elementary Students?

Lessons From an Environmental Engineering Summer Camp for First and Second Graders

By Heather McCullar

With an increased emphasis on engineering in the *Next Generation Science Standards* (*NGSS*; NGSS Lead States 2013), many elementary teachers are left wondering what this might look like at the elementary level. "How do I teach elementary students to think and learn like engineers?" At our school, students get to experience engineering on a regular basis through the school's STEM focus. The school is in its third year of the STEM program in which the mission is "Learning Through Discovery, Leading With Character." As a special part of the program, the school hosted a five-day engineering camp over the summer. The goal was to introduce first- through fifth-grade students to engineering and to address the engineering design standards found in the *NGSS*.

The camp lessons were adapted from the Engineering is Elementary (EiE) curriculum units *Catching the Wind: Designing Windmills* (Museum of Science, Boston, 2013a) and *Now You're Cooking: Designing Solar Ovens* (Museum of Science, Boston, 2013b). This chapter focuses on the lessons, taught in a 5E format (Bybee et al. 2006), and student experiences in the first- and

second-grade room, which had boys and girls from a range of schools across the district.

Engage

As the teacher, I wanted my first- and second-grade students to leave camp with a few basic understandings. First, I wanted students to begin to develop definitions of *engineer* and *technology*. I asked the students to draw and share a picture of an engineer. Bodzin and Gehringer (2001) suggest using drawings of scientists to challenge traditional stereotype perspectives on scientists. Many students drew pictures of men fixing objects or driving vehicles, which indicated they did not have much background knowledge about engineers.

The drawings gave me a starting point for our discussion on engineers. It was clear we needed to develop an understanding over the week of what an engineer does. The camp helped us do this as we explored engineering fields and completed engineering challenges. After the drawings, the students completed a checklist called "What Is Technology?" They selected all the objects they thought were technology. Students were then paired up and given a bag with an object inside (e.g., stapler, pencil, calculator) to observe and sketch, and then they were asked to write what problem the object solved. During our group discussion of the task, students began to notice that all the objects were examples of technology. I asked them to think about their first drawings and if any of their engineers were using technology. Many students said no. I explained that we would revisit and refine our definition of an engineer while we explored different aspects of engineering throughout the week. We developed definitions of *technology* and *engineer* and displayed them in the room:

- "Technology is something created to solve a problem."

- "Engineers are people who use math, science, and technology to create a solution to a problem."

Explore

Through our discussions, the students showed they had developed a basic understanding of engineers and technology. My next step was to provide experiences to help the students develop some of the *NGSS* science and engineering practices by learning about the engineering design process (EDP), which includes identifying a problem, asking questions, making observations, gathering information, and designing a solution (NGSS Lead States 2013). The students were challenged to design and build something that can carry a Ping-Pong ball from the top of a zip line string to the bottom in four seconds or less using the following materials: one small plastic or one large foam cup, paper clips, tape, a hole puncher, and one washer.

Students worked in pairs to draw a sketch that illustrated the details of their design, such as shape, material, and size. Creating sketches allowed students to demonstrate their application of the K–2 engineering standards (NGSS Lead States 2013) by sketching the material, shape, and size they felt was best suited for their design. Then, students worked to construct their model and test it using a premade zip line in the back of the classroom. As they tested their designs, I asked them questions and encouraged them to make modifications to improve their design. When talking with the groups, I would challenge them with questions such as, "How could you modify that to make it go faster?" or, "What might be preventing your design from sliding?" We concluded the lesson with students sharing the process they used to complete the engineering design challenge.

To expand on the students' new understanding of engineers, I introduced them to a variety of engineering fields. I designed centers (for full descriptions, see NSTA Connection, p. 128) where students could explore five engineering fields (architectural, aerospace, ocean, civil, and chemical). Students had 15–20 minutes to work as a group and explore each engineering field. I introduced each station by explaining the task and materials available to the students. Each station was also labeled with its engineering field and task. I circulated among the groups and asked questions and observed their work. Students learned about architectural engineering by designing a "skyscraper" with blocks. They explored aerospace engineering by designing a parachute that would keep a paper clip from crashing down to the floor. For ocean engineering, they designed a submersible to collect a magnet from the bottom of a bucket of water and to resurface. The groups studied civil engineering by constructing a dam to hold back water using various tile pieces. They explored chemical engineering by combining ingredients to make homemade "slime." At the end of the rotations, we gathered as a group and reviewed the engineering fields and the tasks they had completed.

Explain

The environmental engineering focus for this grade level at camp was wind power and windmills. This project allowed the students to apply their knowledge of the EDP and apply the science and engineering practices outlined in the *NGSS* to a new task (NGSS Lead States 2013). I introduced the concept by reading *Leif Catches the Wind*, a story provided in the EiE curriculum (Museum of Science, Boston, 2013a). In the story, students are introduced to the main characters, Leif and Dana, who decide to design something to put oxygen back into Dana's pond so the fish could breathe easier. This design challenge required Leif and Dana to learn about the power of wind and how it can be used to run a windmill.

After the story, I gave the students a new challenge: design a sail blade that would catch enough wind to move a boat from one side of our testing station to the other. The goal was to have students explore the effect of wind on various sail designs with different shapes, sizes, and materials. I wanted them to begin to see that by testing various materials, they could determine which was best suited for their sail design. I asked them to choose from the following materials for their sail design: a small white notecard, a piece of wax paper, a piece of felt, a piece of aluminum foil, tissue paper, a small paper cup, tape, one regular piece of copy paper, and a craft stick to use as the mast. As a group, we developed a properties chart of each material that helped the students determine which one to use when they designed their sail. Before sending students to begin their work, I displayed the fan that would be our source of wind and discussed the safety expectations for the testing station (walking safely around the fan and testing

station, standing back while others were testing, only allowing adults to touch the fan). The students designed their sails, tested them on our model boat, and made modifications to improve the designs. To end the lesson, students shared their sail designs and the results of their tests and modifications.

Students constructing parachutes at the aerospace engineering station

The last camp challenge was to design a windmill that would spin and lift a cup of washers using a piece of string. As a group, we reviewed what we had learned about our sail designs that would help us design our windmills. Students were given the following materials: one half-gallon juice container, rocks (to give the carton weight and hold in place), two to four craft sticks for the blades, sail materials from the previous day to attach to the blades, one small paper cup, one dowel rod, one precut string, tape, and scissors. Students began designing, testing, and modifying their windmills. Any students who designed a windmill that lifted one washer were then challenged to make modifications to lift additional washers. Not all students were able to design a windmill that could lift washers, but all students successfully designed windmills with blades that turned. The students' different levels of success were an important part of the process. We discussed the importance of making modifications and noted that everyone had to make changes even if their windmill worked the first time. The varying levels of success with the windmills also taught students the important lesson that all engineers encounter problems and setbacks, which is simply part of the EDP.

Extend

Our camp ended with a class celebration with parents coming to see what we had learned. I asked each student to share about one of the things we had learned about over the course of the week. The students shared our engineering field stations, our zip line challenge, our sail challenge, and our windmill designs. They took ownership in their learning and now played the role of teacher for their parents. Many parents told me their child was now talking about engineering challenges at home, too. By the end of the camp, the students were the engineering experts and they shared with anyone who dared to stop by their station.

Evaluate

It was important to measure the amount of learning that took place in this first year of the camp. I assessed the students' learning formally and informally during each phase of the week's learning cycle. During the Engage phase, I analyzed the student pictures of engineers and student responses on the "What Is Technology?" assessment provided in the EiE curriculum (Boston Museum of Science 2013). I monitored student understanding during the Explore phase by conferencing with groups as they worked on their zip line design, observing students as they developed their plan, and looking at their science notebook entries. The students had their own engineering notebooks in which they drew the pictures of the engineers, sketched their zip line and sail designs, wrote observations about the engineering field stations, and more. I could quickly monitor each student's understanding by glancing at their engineering notebooks during the week. To evaluate the explain phase, I visited students as they designed, tested, and improved their windmill designs, and the conversations and observations gave me insights into what skills and concepts they had learned.

I used two assessments from the EiE curriculum to evaluate student learning pre- and post-camp for the extend phase. The "What Is Technology?" assessment provided evidence of how students' perspectives on technology changed as a result of the camp. By the end of the week, most students indicated technology was anything developed to solve a problem and showed they realized it did not have to be electronic. The second assessment, *What Does an Engineer Do?*, asked students to indicate which of the listed jobs were ones that would be done by an engineer. Positive results from the posttest and from the students' ability to independently share their engineering stations at our final showcase showed students were beginning to apply their understanding of engineering to different tasks they might perform.

Conclusion

Not many elementary students get the opportunity to learn about engineers, much less engage in engineering activities. At our school, engineering is seen as critical to elementary education. The lessons learned in the camp were just the beginning. The excitement seen in both the girls and boys at this camp indicates engineering is an area in which all students can learn and succeed. The *NGSS* includes engineering standards at the elementary level; therefore, we must include this component in our classroom instruction. I strongly believe integrating engineering aspects into daily classroom instruction is essential for elementary teachers. By teaching our students how to identify problems, plan and design possible solutions, make modifications, and share their designs, we are preparing them to be successful in any career path they might choose. Better yet, we are teaching our students how to learn through discovery and to engage in experiences where they can apply what they have learned about engineering and the EDP.

Connecting to the *Next Generation Science Standards*

The materials, lessons, and activities outlined in this chapter are just one step toward reaching the performance expectations listed below. Additional supporting materials, lessons, and activities will be required.

2PS-1 Matter and Its Interactions *www.nextgenscience.org/2ps1-matter-interactions*	Connections to Classroom Activity
Performance Expectation	
2-PS1-2: Analyze data obtained from testing different materials to determine which materials have the properties that are best suited for an intended purpose	Built models to determine what shape, size, and material was best suited for a windmill sail or blade
Science and Engineering Practices	
Developing and Using Models	Designed, built, and gathered information about a sail
Planning and Carrying Out Investigations	Planned, designed, tested, and modified a zip line carrier
Constructing Explanations and Designing Solutions	Designed, built, and explained a model windmill blade to exhibit and test effectiveness of chosen material
Obtaining, Evaluating, and Communicating Information	Communicated details to partner
Disciplinary Core Ideas	
PS1.A: Structures and Properties of Matter • Different properties are suited to different purposes.	Designed a windmill using the materials, size, and shape of blade best suited to catch the wind and lift washers
K-2 ETS1.B: Developing Possible Solutions • Designs can be conveyed through sketches, drawings, or physical models.	Created sketches and a model of blade and windmill designs
Crosscutting Concept	
Structure and Function	Evaluated the effectiveness of the chosen sail shape, size, and material for the given task

Source: NGSS Lead States 2013.

References

Bodzin, A., and M. Gehringer. 2001. Breaking science stereotypes. *Science and Children* 38 (4): 36–41.

Bybee, R. W., J. A. Taylor, A. Gardner, P. Van Scotter, J. C. Powell, A. Westbrook, and N. Landes. 2006. *The BSCS 5E Instructional Model: Origins and effectiveness.* Colorado Springs, CO: BSCS.

Museum of Science, Boston. 2013a. *Catching the wind: Designing windmills.* Boston, MA: Engineering is Elementary.

Museum of Science, Boston. 2013b. *Now you're cooking: Designing solar ovens.* Boston, MA: Engineering is Elementary.

NGSS Lead States. 2013. *Next Generation Science Standards: For states, by states.* Washington, DC: National Academies Press. *www.nextgenscience.org/ next-generation-science-standards.*

NSTA Connection

Download a description of the centers at *www.nsta.org/sc1503*.

The STEM of Inquiry

By Peggy Ashbrook

Many early childhood science lessons have children making observations of an organism or phenomena and documenting their observations (science). Counting and taking measurements may be involved (mathematics), and students may use tools to observe and document (technology). However, engineering concepts are often left out when they could easily be included. Planning to include an engineering process as part of an investigation can help us see how engineering can be an important part of the investigation—not a short activity tacked on to satisfy the "E" in STEM.

A science and engineering investigation may become an inquiry over time as children ask many related questions, investigate, collect data, and talk about their ideas. Deep exploration of concepts over time will allow students to use science and engineering practices and build toward mastery as their understanding matures. Children learn more in-depth when they—with teacher support—are able to pursue areas of interest (Moomaw 2013). Just as it is likely that additional questions to investigate will come up as part of a science inquiry, the engineering

design process is iterative—the process can continue as students continue to ask, imagine, plan, create, improve, and so on.

The Engineering is Elementary program (see Internet Resource, p. 132) has resources for elementary students that preK teachers can use to learn about the engineering design process. While reading about how older children investigate an engineering design problem (Bricker 2009), consider what is developmentally appropriate for your students. Engineering design disciplinary core ideas for kindergarten through grade 2 state that students who demonstrate understanding can "ask questions, make observations, and gather information about a situation people want to change to define a simple problem that can be solved through the development of a new or improved object or tool" and "develop a simple sketch, drawing, or physical model to illustrate how the shape of an object helps it function as needed to solve a given problem" (NGSS Lead States 2013).

Plan for an activity that can inspire you and your students to work on a question or problem over a long period of time—enough time that questions can be fully explored to the satisfaction of young children. Children take on many tasks that involve solving a problem—building a tower of small blocks when all the large blocks have been used by others, for example. Some of their self-appointed problem-solving tasks may suggest a topic for future investigation by the entire class.

Heavy Lifting

Objective

Children will investigate how to carry and transport a heavy object including designing and using a tool.

Procedure

1. Read through this or another activity to note how science, technology, engineering, and mathematical concepts are involved. Plan to include this activity as part of a larger inquiry or project, such as one about simple machines, motion, or transportation.

2. Introduce the heavy object to your class by having students pass it around the circle. Challenge them to describe it so you can send the description home to their families. The name of the object will not be on the note; family members can have fun guessing what the object is. Remind the children that the name is going to be a secret. Children may not be able to describe more than color, shape, and other familiar attributes. Prompt them to also describe the texture, the smell (if any), and the weight and size (relative or measured), and ask them to observe the motion when gently pushed.

3. As the children work, observe aloud that they are working as scientists when you see them using their senses and describing, and that they are using math when they measure. Record, or have students record, the descriptions and measurements with text and drawings, according to their development.

4. Introduce another challenge: to move the object safely (no breakage, bruising, spillage, or scrapes) from one place to another (locations chosen by you) without simply carrying it in their hands. Comment that engineers draw their ideas for solving problems such as this one. Through drawing, children may consider additional design ideas, which is a habit of designing that can lead to success.

5. Provide a variety of materials that students can use to create a carrier—to small groups or at a center. Assist students in cutting or tying materials, support their thinking about what is or is not working (without telling them how to do it), and prompt them to redesign and rebuild if they appear frustrated. Ask students to explain their choice of materials to support their consideration of the properties of the materials and any alternatives.

6. Children can show that their idea successfully moves the heavy object by demonstrating to the small group or by documenting with drawings, photographs, or video.

If a design does not successfully move the object, ask, "Were there any problems moving the object?" and "What one feature or part of your carrier could you change to make it work?" Noticing a problem with a design and changing one thing for each new test will help students understand that redesigning is expected and can determine what single change was needed to make the carrier work. Review the children's drawings or photographs and talk about how each carrier design transported the object safely. Write a note to send home using the students' descriptions. Read a book that extends the learning about making observations; measuring, making, or improving a tool; or solving a problem (see NSTA Connection, p. 132).

Materials

- Heavy objects (e.g., a pumpkin, large water balloon, bag of sand)—1 per small group

- Magnifiers, measuring tools (standard or nonstandard), weighing scale, camera (optional)

- Writing and drawing materials

- Various materials and tools familiar to the class that can be used as part of a system for transporting the heavy object (e.g., scarves, boards, baskets, cardboard boxes, scooter boards, toy vehicles, tape and scissors, short pieces of rope or yarn)

References

Bricker, M. 2009. Plants on the move. *Science and Children* 46 (6): 24–28.

Moomaw, S. 2013. *Teaching STEM in the early years: Activities for integrating science, technology, engineering, and mathematics.* St. Paul, MN: Redleaf Press.

NGSS Lead States. 2013. *Next Generation Science Standards: For states, by states.* Washington, DC: National Academies Press. *www.nextgenscience.org/ next-generation-science-standards.*

Internet Resource

Engineering is Elementary
www.eie.org

NSTA Connection

For STEM reading recommendations, visit **www.nsta.org/SC1310.**

Printing the Playground

Early Childhood Students Design a Piece of Playground Equipment Using a 3-D Printer

By Stephanie Wendt and Jeremy Wendt

With grant funds in play, teachers at Prescott South Elementary School in Cookeville, Tennessee, were looking for innovative ways to engage students in new and exciting STEM experiences. One outcome was the purchase of a MakerBot three-dimensional (3-D) printer. The printer provided opportunities to complete previously inconceivable projects. Teachers submitted ideas about ways to interest students in 3-D modeling processes while meeting standards and school objectives. The school STEM advisory committee posed a vision that challenged both teachers and students. Kindergarten and first-grade students were charged with designing new playground equipment to create a solution to a problem: the school playground needed additional equipment. Students jumped at the opportunity to solve an issue they were dealing with daily.

This project engaged students in science and engineering practices as described in the *Next Generation Science Standards* (*NGSS*; NGSS Lead States 2013). Several tech-savvy teachers (and two middle school students) were trained to use the printer and 3-D modeling programs such as

SketchUp—chosen as the primary implementation software because of its ease of use and free access for educators. In turn, they became facilitators and troubleshooters for student learning projects.

The playground project began with a week of interdisciplinary planning by teachers across the K–1 classes, followed by the weeklong implementation. For the culminating project, students presented the principal with a 3-D printed model of the playground equipment selected from their designs. In turn, the principal ordered the equipment for the school. Throughout the authentic learning experience, the students engaged in science and engineering practices. They asked questions and defined problems, developed and used models, planned and carried out investigations, and constructed explanations and designed solutions. The interdisciplinary project incorporated literacy, science, mathematics, and physical education. In addition, throughout the unit, teachers reported an increased level of engagement.

It is important to note that although this engineering design challenge culminated with the use of a 3-D printer, the process is still valuable for students who do not have access to this technology. If the support structure for 3-D printing is not in place in the school district, then local universities and 3-D printer companies are other possible resources for providing training and support. For this project, the 3-D printer was obtained through a STEM grant, but it could have been purchased through a variety of sources. The school PTO, *www.donorschoose.org*, and *www. grantwrangler.com* are all reasonable avenues for funding a 3-D printer (see Internet Resources, p. 140). Collaboration with local businesses and universities could provide opportunities for 3-D printing as well. A business or university with a 3-D printer could easily receive the file and print the object for the school. Video conferencing could even be used to allow students to view the process and speak to the operator as the process is happening. 3-D printing software is rapidly evolving and improving. Based on the experiences of the local school and university discussed in this chapter, companies are generally very responsive to requests for assistance and are eager to expand their uses for implementation in the K–12 setting. The 5E Model, developed by BSCS (Bybee et al. 2006), was one of the foundational approaches for the project.

Engage

To generate excitement about the project, the principal wrote a letter to the kindergarten and first-grade classes requesting their help to design new playground equipment for their school. After this introduction, the students were shown videos, pictures, and examples of children on various types of playground equipment to help them brainstorm ideas. First individually and then in small groups, students journaled about specific equipment and activities they desired. They described their ideas creatively in writing and with annotated student drawings (Figure 17.1). The annotations included

comments such as, "It needs to be wider at the bottom to hold more people," "The bars need to be close together for us to reach," "If we all stand next to each other, how big does it need to be to hold us?" and "Should we use plastic or metal on our creation? We want it to last a really long time!"

FIGURE 17.1

Annotated Student Drawing

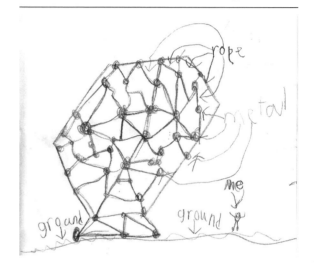

3-D Printing

What exactly is a 3-D printer and how does it work? The small-scale 3-D printer has evolved from large-scale "additive manufacturing." For public sector use, this translates to adding materials in layers to construct a solid, 3-D object. Essentially, once you've created or downloaded a 3-D digital model, you send it to the printer through a USB or wirelessly. The device then heats and melts string-shaped plastic from a spool in layers onto a platform to build the model. The device makes passes similar to an inkjet printer, depositing the plastic material. The process can take a few minutes or several hours to complete, but the possibilities are endless.

According to Page Keeley (2008), "Annotated Student Drawings are student-made, labeled illustrations that visually represent and describe students' thinking about a scientific concept." This concept, in addition to Think-Pair-Share (see more in the explain stage), is one of the 75 science formative assessment classroom techniques from Keeley's book. Students were engrossed in the ideas of equipment design and engineering, while employing higher order thinking skills. In this stage of the 5E model, students were introduced to initial steps of the engineering design process (Lachapelle and Cunningham 2007) by asking questions and defining problems while imagining equipment on which they wanted to play.

In the Explore stage, students use manipulatives to develop and build their playground models.

Explore

The teachers facilitated and moderated discussions to help the students visualize the complex processes that architects and engineers follow to make project ideas a reality. The NGSS science and engineering practice Developing and Using Models states, "In engineering, models may be used to analyze a system to see where or under what conditions flaws might develop, or to test possible solutions to a problem. Models can also be used to visualize and refine a design, to communicate a design's features to others, and as prototypes for testing design performance" (NGSS Lead States 2013). In this stage, the students began drawing and designing more complex variations of their ideas. Throughout the 5E process, and especially during this stage, teachers incorporated mini-lessons that addressed students' questions on topics such as geometric shapes, addition, subtraction, and other subjects as needed. For example, CCSS.MATH.CONTENT.K.G.B.4: Analyze, Compare, Create, and Compose Shapes (NGAC and CCSSO 2010) was addressed in a mini-lesson with pattern blocks and geometric shaped manipulatives. Students described two-dimensional and 3-D shapes' similarities, differences, and attributes in small groups and with the class. In first grade, teachers and students addressed CCSS.MATH.CONTENT.1.G.A.1: Reason with Shapes and Their Attributes (NGAC and CCSSO 2010) while constructing triangles, rectangles, and other geometric shapes out of classroom materials and manipulatives. Students defined the specific attributes of shapes individually and in their small groups. Teachers incorporated flipcharts and internet resources on the interactive white board as part of whole-class instruction.

Once students demonstrated their understanding of the basic geometric shapes and concepts, they were given the opportunity to take all of the individual lesson components and compile them into a working model. Students synthesized all of the information from the challenges, lessons, and processes into a physical reality. This was accomplished by using manipulatives such as gumdrops, marshmallows, and toothpicks to develop physical models of the playground equipment sketches. During their investigation, students worked like engineers to explore different heights and bases for their models. They experimented to learn which designs were strongest and discussed their ideas in their groups and with the teacher.

Explain

With guiding questions from the teachers, the students assessed the current playground equipment. They brainstormed innovative possibilities for improving the existing playground, including the best materials, structural designs, safety considerations, and costs for their ideas. In this phase of the project, students continued planning following the Think-Pair-Share model. According to Keeley (2008), Think-Pair-Share combines thinking with communication. The teacher poses a question and then gives students time to think about the question individually. Students are then paired with a partner to discuss and share their ideas in small-group or whole-class discussions.

Each group had one member assigned as the group note-taker to summarize their collective ideas. They determined their current playground's strengths and weaknesses, working through a number of possibilities concerning a variety of equipment. The students' progress was monitored and prompted by teachers through guiding questions, such as the following: Is wood, metal, or plastic the best material for your design? How many children can play on your model at one time? Is your model too tall for kindergarten and first-grade students? Throughout this stage, students constructed explanations and designed solutions to help address several aspects of playground design: the number of students that could play on one piece of equipment at a time, safety issues, materials needed, appropriate size and type of equipment based on student age, aesthetic properties, and durability. The teachers guided students to carefully consider each of those points during the planning for their project, keeping the end product in mind.

Interactive Resources for 3-D Printing

For lower grades, students can use design apps for the iPad that will print to a 3-D printer. Different 3-D printers have product-specific software, but general design software files can be exported to individual printers. Blokify is an app that lets users create "blok"-based models through a guided building experience or free-form. Users then send the completed model to the 3-D printer for creation. Several websites, such as *www.thingiverse.com*, have hundreds of models available for download that can be transferred to a 3-D printer for creation.

SketchUp is a 3-D modeling program that is used by architects, engineers, interior designers, and many other professionals. The software has an online, open-source repository called 3-D Warehouse, with free, downloadable models. A freeware version, SketchUp Make, is available to the public. Educators can apply to receive the full-featured version of the software, SketchUp Pro, free of charge (see Internet Resources, p. 140).

Elaborate

After having small-group discussion about the models, groups presented their designs to the class. Students were asked to explain the thoughts, ideas, and processes behind their designs. The thought processes of the students were observed in the following commments made during presentations:

- "If we build it really tall, I could probably see all of my friends from the top."

- "If we build it out of metal, it will last longer than something made out of wood."

- "We can't all play on one thing at the same time, because it would break if we were all on it."

This step in the process allowed students to experience engineering design in an authentic manner. Following their presentations, a teacher who had attended the 3-D printing workshop transferred ideas from students' plans to SketchUp, enabling students to view their conceptual designs in the digital environment. Student designs were selected and constructed based on the most realistic and viable options for new equipment. These samples were passed around the classrooms for the students to analyze and view. From the many geometric designs and shapes presented, the principal selected three realistic choices. The students voted on a variation of an *icosahedron* as their final choice.

Evaluate

Formative assessment was used throughout the project. Teachers kept students on task by asking "Are you meeting the objectives of the project?" "Have you stopped and discussed improvements you could make?" and "Are you following directions?" Students were asked to make changes and adaptations to their designs based on group collaboration and feedback from teachers. Assessments included the following:

- Use of tens frames helped assess students' addition and subtraction skills.

- Students were formatively assessed through the use of NCTM's Illuminations interactive website (see Internet Resources, p. 140).

- Students were formatively assessed through the use of the Science Kids'

interactive simulations to determine properties of materials (see Internet Resources, p. 140).

- Students were asked to clearly define the problem orally to the teacher before developing models.

- Using their models and drawings, students successfully communicated solutions with group members and then to the whole class.

A student examines a 3-D printed playground equipment model.

- Students made bar graphs to represent the data collected from student votes

- Students' knowledge of geometric shapes and measurement skills was summatively assessed.

Looking forward to the end result of playground construction and its impact on the school, students were very involved and took ownership of the project. The principal made announcements to the school throughout the process and gave regular updates to the classes. Students remained eager each day to hear about the next step in the project's progress. Students were overheard explaining the mission and excitement of the project and the new playground:

"This is hard being an engineer, but it's fun!" "My brother won't believe that we were the ones who made the playground." The culminating event of the project came when the principal selected student designs to be ordered and purchased for the school playground. A large rope climber with geometric shapes that incorporated the kindergarten and first-grade students' designs was installed in May on the elementary playground. The students and faculty members celebrated their success with an unveiling and dedication.

Reflections

As a first-time project, the planning to implementation period was brief, and more time could be allotted to the hands-on and exploratory model creation. Approximately two weeks were dedicated, but three weeks would have been more accommodating. Additional support and outside perspectives would be beneficial to the project. For example, involvement and collaboration from an engineer or architect would help in understanding the design processes. If present, preservice teachers from a local university could help with the project by increasing the individual attention given to students. Preservice teachers could be prepared for this type of project by replicating the process in the college classroom. They would understand and be able to implement the project in many settings, but having the experience and knowledge would prepare them for the classroom.

Opportunities

K–1 grades are often overlooked when advanced projects are undertaken. The playground project demonstrated that this age group benefited from being part of the process. This foundation of critical thinking and problem solving can be built in the early grades to enable advanced processing and knowledge base in the later grades. In higher grade levels, problems can become increasingly complex. For example, fourth-grade students at the school addressed the problem of noise in the cafeteria by designing sound tile samples on the 3-D printer. The 3-D printer aids in the scale creation of these and many other types of models. Even without access to a 3-D printer, engaging students in this process creates an environment for excitement and opens doors to the future. Teachers and students can use resources available to them such as modeling clay, toothpicks, candy, tinker toys, Legos, and other classroom manipulatives to assist in problem solving. Professionals in STEM careers work to solve similar problems daily. What better time is there to introduce a love for STEM than in childhood?

Connecting to the *Next Generation Science Standards*

The materials, lessons, and activities outlined in this chapter are just one step toward reaching the performance expectations listed below. Additional supporting materials, lessons, and activities will be required.

K-2-ETS1 Engineering Design *www.nextgenscience.org/k-2ets1-engineering-design* **2-PS1 Matter and Its Interactions** *www.nextgenscience.org/2ps1-matter-interactions*	Connections to Classroom Activity
Performance Expectations	
K-2-ETS1-1: Ask questions, make observations, and gather information about a situation people want to change to define a simple problem that can be solved through the development of a new or improved object or tool	Asked questions and engaged in a real-world, relevant problem that affected them directly, resulting in a student-derived solution
K-2-ETS1-2: Develop a simple sketch, drawing, or physical model to illustrate how the shape of an object helps it function as needed to solve a given problem	Developed innovative sketches, modified the drawings as they modified the concepts, and built models to visualize and test design solutions for playground equipment
2-PS1-2: Analyze data obtained from testing different materials to determine which materials have the properties that are best suited for an intended purpose	Tested a variety of materials to determine which would best fit the needs of the project
Science and Engineering Practices	
Developing and Using Models	Constructed models of playground equipment from manipulatives
Asking Questions and Defining Problems	Asked questions and made observations about playground design and materials while analyzing variables such as strength, durability, and safety
Analyzing and Interpreting Data	Explored a range of materials to compare different solutions for the playground equipment design
Disciplinary Core Ideas	
ETS1.A: Defining and Delimiting Engineering Problems • A situation that people want to change or create can be approached as a problem to be solved through engineering.	Presented with the school principal's engineering design challenge to solve the playground equipment problem
ETS1.B: Developing Possible Solutions • Designs can be conveyed through sketches, drawings, or physical models. These representations are useful in communicating ideas for a problem's solutions to other people.	Conveyed equipment designs and explanations through journals, simple sketches, annotated drawings, physical, and 3-D printed models Constructed models of playground equipment from manipulatives

K-2-ETS1 Engineering Design www.nextgenscience.org/k-2ets1-engineering-design **2-PS1 Matter and Its Interactions** www.nextgenscience.org/2ps1-matter-interactions	**Connections to Classroom Activity**
Disciplinary Core Ideas	
ETS1.C: Optimizing the Design Solution • Because there is always more than one possible solution to a problem, it is useful to compare and test designs.	Used Think-Pair-Share to focus discussion and exploration while comparing and testing designs Constructed models of playground equipment from manipulatives
PS1.A: Structure and Properties of Matter • Different properties are suited to different purposes. • A great variety of objects can be built up from a small set of pieces.	Investigated a variety of materials for applicability to the playground structures
Crosscutting Concept	
Structure and Function	Observed and analyzed attributes of geometric shapes and scale models to form concepts of scale size and function

Source: NGSS Lead States 2013.

References

Bybee, R. W., J. A. Taylor, A. Gardner, P. Van Scotter, J. Carlson, A. Westbrook, and N. Landes. 2006. *The BSCS 5E Instructional Model: Origins and effectiveness.* Colorado Springs, CO: BSCS.

Keeley, P. 2008. *Science formative assessment: 75 practical strategies for linking assessment, instruction, and learning.* Thousand Oaks, CA: Corwin Press.

Lachapelle, C. P., and C. M. Cunningham. 2007. Engineering is Elementary: Children's changing understandings of science and engineering. Paper presented at the ASEE Annual Conference and Exposition, Honolulu.

National Governors Association Center for Best Practices and Council of Chief State School Officers (NGAC and CCSSO). 2010. *Common core state standards.* Washington, DC: NGAC and CCSSO.

NGSS Lead States. 2013. *Next Generation Science Standards: For states, by states.* Washington, DC: National Academies Press. *www.nextgenscience.org/next-generation-science-standards.*

Internet Resources

Donors Choose
  www.donorschoose.org

Geometric Solids
  http://illuminations.nctm.org/Activity.aspx?id=3521

Grant Wrangler
  www.grantwrangler.com

MakerBot Thingiverse
  www.thingiverse.com

Measuring Me
  http://illuminations.nctm.org/Lesson.aspx?id=2756

NCTM's Illuminations
  http://illuminations.nctm.org

Science Kids
  www.sciencekids.co.nz/gamesactivities/materialproperties.html

SketchUp
  www.sketchup.com

A House for Chase the Dog
Second-Grade Students Investigate Material Properties

By Meghan E. Marrero, Amanda M. Gunning, and Christina Buonamano

From a young age, children encounter different materials and learn color, hardness, texture, and shape. Focusing on observable properties is an engaging way to introduce young children to matter. In this investigation, students use observations and engineering design to decide which material would make the best roof for a doghouse. We used the 5E model (BSCS and IBM 1989) to create an engaging inquiry-based activity to meet standards and make real-life connections to physical science content. Our second-grade students enjoyed the activity and came to understand how physical properties can determine how a material is used.

Material properties are an important foundational piece of physical science content. As students advance, these early connections will be built on in chemistry, Earth science, and physics. At this age, students are learning that materials are suited for different purposes because of their properties. The children should be able to analyze data related to properties and sort materials based on this analysis. Some material properties appropriate for this age group are strength, flexibility, hardness, texture, and absorbency (NGSS

Lead States 2013). Early exploration of properties and classification helps support continued engineering work in upper elementary grades (Lachapelle et al. 2012).

Lesson Objectives

As two graduate education professors and one elementary school teacher, we have used this activity with preK and elementary students, varying the amount of support provided as was developmentally appropriate. This chapter describes how we implemented the activity with a group of primary students in one class period. Our students were introduced to the *Next Generation Science Standards* (*NGSS*) model of the engineering design process (EDP) and given different materials to test to determine the best solution for creating the roof of a doghouse to withstand rain. We chose to model activities with students first, giving them the opportunity to experience the appropriate science and engineering practices but still allowing us to finish the activity within one class period.

In the end, students were able to (a) draw on personal experience and prior knowledge by explaining that roofs function to protect humans and pets from the elements, particularly precipitation; (b) compare how different materials stand up against rain; (c) use observable data to compare roof materials; and (d) decide which roof material is best on the basis of evidence from their engineering investigation.

Engage

Where do animals and people go when it rains? We used this question as a starting point for discussion leading into the activity. In our class, students said they went inside their homes, cars, and school. From there, we asked why it is important to go to those places (to stay dry) and how those places help you (or animals) stay dry in the rain. Children quickly realized the roofs of all of those places keep the rain out.

Next, we showed the class a picture of a classmate's new dog and asked whether any students have a dog at home and whether their dogs like to be outside. We explained to students that we wanted their help, and we read the following scenario:

> Chase the dog loves being outside. Although he has a family and a warm, dry house, he prefers to hang out in the backyard, no matter the weather. Chase has two loving kids in his house, and they are concerned for their pet. The kids have decided that the best way to keep Chase both happy and safe is to design the perfect doghouse to protect him from the elements. Your task is to begin testing roof designs and to answer the following question: What is the best material to use for the roof of Chase's doghouse to keep him dry in the rain?

We also introduced students to an example of EDP and shared the EDP graphic from the *NGSS*

(Figure 18.1), discussing each step of the process and explaining that engineers design, build, and test solutions to problems. We asked whether any students know an engineer and what engineers do. Discussing engineering generally helps explain the integral role engineers play in our designed world.

FIGURE 18.1

NGSS EDP for Young Learners

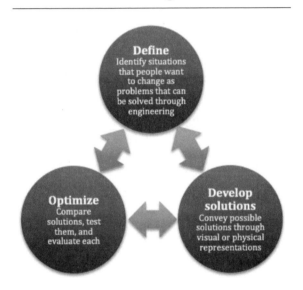

Explore

As with all hands-on activities with young children, this activity should be modeled before materials are distributed to students. It is important to always discuss safety with students before beginning any hands-on exploration. Be sure all milk cartons are cleaned out and prepared for student work. Start by showing the milk carton "doghouse" base and explaining it is a *model*. We introduce the idea that scientists use models all the time to learn more about what is being studied. For example, you may talk about past activities you have done with students, such as modeling friction by driving toy cars over different "roadways" (e.g., rug vs. tile). Explain that engineers use models to test their designs. For this activity, instead of testing materials on a big doghouse in someone's backyard, we use the milk carton base. Ask students questions such as, "Why do you think it is a good idea to use a model?" Show the different roof materials to the class and discuss, "What might make a good roof material? What might we observe happen to a poor roof in the 'rain' (water spray)?" Record discussion comments on the board for visual learners.

Model for students how to lay the squares of material on top of the "house" one at a time and use the rubber band to secure it, if needed, by stretching it around the mouth of the milk carton. Note that students may choose to design their roofs in different ways; these designs can be a basis for later discussion (e.g., some students may choose to crease the materials to make a peaked roof). Once the testing process has been modeled for students, show the predict, observe, explain (POE) chart (see NSTA Connection, p. 148). Explain that for each material they test, students will need to predict what will happen when they sprinkle water on it before they actually do so. It is a good idea to model

Materials

The following materials are required for each group of three students:

- Precut quart-sized milk carton
- Cups
- Water
- Spray bottle filled with water
- Eye droppers (1–3 per group)
- Thin, large rubber bands
- Aluminum baking trays

The following required materials should be cut to 10 in. square:

- Wax paper
- Poster paper
- Cardboard
- Aluminum foil
- Construction paper
- Cloth (e.g., old T-shirts)
- Shower curtain

Other required materials include the following:

- Sharp scissors (for adult use to cut cardboard and milk cartons)
- Children's safety scissors (to cut other materials, if desired)

Preparation

For each of the materials, cut a square for each group that is approximately 10 in. by 10 in. Students in grade 2 can measure and cut materials on their own, as an additional learning activity incorporating math. Care should be used if students are cutting materials themselves, and only adults should be cutting the thicker materials, such as cardboard or the milk cartons.

Save enough empty, quart-sized milk cartons for the class and cut each milk carton all the way around, about 3 inches from the bottom. This makes the bottom of a "house."

Create a setup for each group in the baking trays, where you place in each a cut off milk carton, eye dropper, cup (to be filled with water for the eye dropper), rubber band, and spray bottle.

what is expected and have the students reiterate the directions before they begin. We also found it useful to have students make predictions for each of their roof materials before we handed out any water. If working in groups is new to your students, be sure to model how to work collaboratively and to continue to discuss, write, or illustrate thoughts until they are told to stop.

Students should be divided into groups of three, if possible. We created our groups by considering the level of each child's cognitive stage of development. We addressed the needs of students who are at an introductory level as well as those who are ready for more in-depth, higher level thinking. This grouping allowed for varied levels of scaffolding to take place during the lesson. Tiering lessons and differentiation occurs on a daily basis, so children were unaware of the grouping methods while working together at their level of readiness. Higher performing groups were given roof materials that had similar properties, or were all generally good roofs (e.g., cardboard, shower curtain, foil). In this way, higher performing groups would have to distinguish observations and decide which material was best based on its properties. Groups of lower performing students were given roofing material with observable differences, such as construction paper, cloth, and a shower curtain. You may also choose to differentiate groups according to learning styles or any other way that you feel will benefit your students, while providing them with the same material to complete the lesson.

Once students had their materials (but not any water), they examined the possible roofs and listed each option under the first column on their group POE chart (Figure 18.2). Students then predicted what they thought would happen to each material in the rain and recorded their predictions in the second column. After much anticipation, students were given the sprayers and cups with water and droppers. Students should be reminded not to spray one another. Also be mindful of slip hazards from wet floors. Remind students to use their eyes to see the roof and their fingers to touch the roof as part of observing. They can also look inside the house. Observations should be recorded in the third column of the POE chart. Students might also consider drawing their observations.

FIGURE 18.2

Sample POE Chart

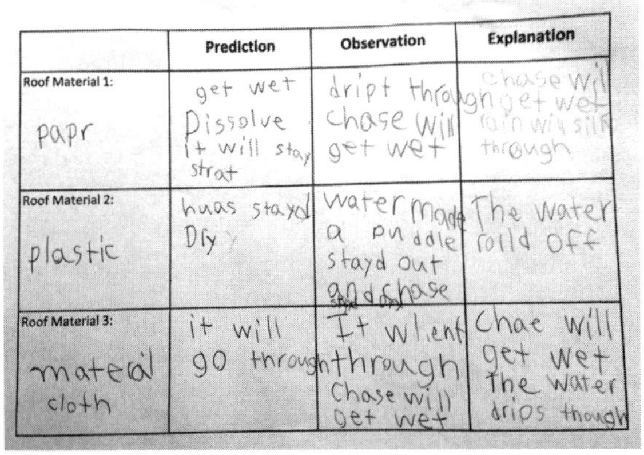

	Prediction	Observation	Explanation
Roof Material 1: papr	get wet Dissolve It will stay strat	dript through Chase will get wet	chase will get wet rain will slip through
Roof Material 2: plastic	huas stayd Dry	water made a puddle stayd out and chase	The water rolld off
Roof Material 3: materal cloth	it will go through	It wlent through Chase will get wet	Chae will get wet The water drips though

Explain

After each observation is recorded, students should fill in the Explain column to decide whether a given material has properties that would make it a good roof and why. It is important for students to provide evidence for their assertions. After the experiment, teams of students brought their POE sheets to the rug and shared their findings, discussing which material they thought was the best for the doghouse roof and sharing their supporting evidence. The *NGSS* remind us that we must ask students to explain the solutions they have developed, "Asking students to demonstrate their own explanations … [of] models they have developed, engages them in an essential part of the process by which conceptual change can occur" (NRC 2012, p. 68). We

TABLE 18.1

Sample Teacher Questions and Student Answers

Teacher Prompt	Sample Student Responses
Which material did the best job holding out the rain? Why?	Students discussed how the curtain was best because the water slid off of it, and the foil was good because the water didn't get in the house.
How did you know if the material would make an effective roof?	"Chase will get wet. Rain will sink through." "When the wax paper got wet my finger went through." "Paper ripped. It is not a good roof."
What happened when you used a piece of shower curtain?	"The water stayed on the roof and did not go in there." "The water rolled off."
How did the cloth perform in the experiment?	"Chase will get wet the water drips through."
How well did your predictions match up with your observations? Were you surprised by anything you found?	Students in one group discussed that they noticed the construction paper worked as long as it didn't rain hard. When it rained hard, it got soft and you could make a hole in it, which prompted us to discuss the importance of the roof being able to sustain several rainstorms.

helped students make connections between the structure and properties of the materials (e.g., the wax paper had a coating that caused water to bead up). We used questions similar to those in Table 18.1 to spark discussion. We included sample student responses taken from their POE sheets' Explain column.

Stress to students that they are using scientific observations (i.e., data) to support their ideas. Scientists and engineers must rely on data to make decisions—in this case, about which type of roof would be the best. Material properties of the roofs dictate their performance (e.g., smooth, non-porous materials, such as the shower curtain and foil, will allow water to roll off). Other materials that are porous, such as the cloth, will get wet and eventually may allow water to drip through. Direct the teams to share their findings. Because they each tested different materials, ask

them how they can make a decision about the best roof material. Explain that there is not necessarily one right answer, but that students should use evidence, which is their observational data, to support the solution they designed. This connects directly with the *NGSS* science and engineering practice of Constructing Explanations (for science) and Designing Solutions (for engineering).

Ask students why some materials worked better than others for this purpose. Students should be able to explain that different materials have different characteristics, or properties. Some of the materials, for example, have the property of absorbing or repelling water. We asked students, "What are some properties that are important for designing a roof? Answers included, "Being strong; Being firm; Preventing water from going through." Discuss the students' findings in light of the concept of properties. Record the properties students share and explain properties are a way of describing a material or matter, in general. Students should come to understand that different material properties can make that material better suited for different functions, highlighting the *NGSS* crosscutting concept of Structure and Function.

Through questioning, help students to understand that the properties of different materials make them suitable for different purposes. Ask, "Why is cloth better than a shower curtain for a T-shirt?" (The fabric is much more "breathable" and comfortable.) If applicable, the discussion may also shift to the shapes in which students designed their roofs. Some student teams may have designed their roof to be more sloped, whereas others flattened. Use these differences as an opportunity to further discuss structure and function, asking students to share how effective the different shapes were at keeping the rain out of the house and at ensuring the roof is strong.

Elaborate

Have students describe the materials from which houses or buildings are made in your area. Ask students whether they have traveled to different places and whether the homes looked similar or different. Ask whether they think that houses are made from the same materials everywhere and why. Show students some photographs of different types of homes (e.g., a mud hut in Africa, Adobe house in the American Southwest, yurt in Mongolia). Ask students to share why they think the homes are made of different materials (i.e., ask them to consider the materials available in different areas and also different climates). Ask students the following questions: "What kinds of roofs do these houses have? Why are these features important for these roofs?"

To further explore material properties, see Internet Resource (p. 148) for a website to test metal, glass, rubber, paper, and fabric for transparency, flexibility, strength, and water resistance. The site reads the text out loud if the speaker icon is clicked, which is a great support for English language learners. It also features a quiz students may try. This application can be explored as a whole class or by individual or small groups of students, but the teacher should model it first.

Evaluate and Extensions

Opportunities for formative assessments are present throughout this activity. Begin in the whole-class portion of the lesson by drawing on students' prior understandings of shelter and its function. While working on the activity, observe how student teams are testing the materials, and note student conversation and the observations and explanations cited on their worksheets to assess how they use engineering practices. During the class discussion, each team should share ideas and support them with evidence, allowing the teacher to assess how students are constructing their explanations and designing solutions.

As a homework assignment and extension of the activity, ask students to think about the materials they tested in class. Using those materials, students could design a coat for Chase to wear when he is in the yard, in case of rain. Students should find a design team partner at home—a parent, sibling, or other caregiver—to help with the assignment. Students should explain the roof testing that was done in school and how the different materials performed in the rain. Using this information, they should determine which material or materials would be the best to create a coat for Chase. Students should specifically include the material properties of the items used in their design and explain why they were chosen (see NSTA Connection, p. 148).

Connecting to the *Next Generation Science Standards*

The materials, lessons, and activities outlined in this chapter are just one step toward reaching the performance expectations listed below. Additional supporting materials, lessons, and activities will be required.

2-PS1-2 Matter and Its Interactions www.nextgenscience.org/2-ps1-2-matter-and-its-interactions **K-2-ETS1-3 Engineering Design** www.nextgenscience.org/k-2-ets1-3-engineering-design	Connections to Classroom Activity
Performance Expectations	
2-PS1-2: Analyze data obtained from testing different materials to determine which materials are best suited for an intended purpose	Tested various materials and used observations to decide which materials were best suited as a roof to withstand rainfall
K-2-ETS1-3: Analyze data from tests of two objects designed to solve the same problem to compare the strengths and weaknesses of how each performs	Compared different roof materials in terms of how they perform under model rainfall and analyzed the strengths and weaknesses of each
Science and Engineering Practices	
Analyzing and Interpreting Data	Collected qualitative data in a chart while testing different materials under spray bottle "rainfall" Discussed observations with group members Interpreted data to determine whether each roof material was effective Explored an online application that tests different materials and properties
Constructing Explanations and Designing Solutions	Explained why roofs are important and what features are needed to make a good roof Designed solutions to the problem of building the best roof for a doghouse to withstand rainfall
Disciplinary Core Ideas	
PS1.A: Structure and Properties of Matter • Different properties are suited to different purposes.	Tested materials' suitability for use as roofing material Discussed properties of materials observed in the activity Observed various home designs and discussed properties of different materials used to build them Discussed which properties were better suited for a shower curtain or T-shirt

2-PS1-2 Matter and Its Interactions www.nextgenscience.org/2-ps1-2-matter-and-its-interactions K-2-ETS1-3 Engineering Design www.nextgenscience.org/k-2-ets1-3-engineering-design	**Connections to Classroom Activity**
Disciplinary Core Ideas	
ETS1.C: Optimizing the Design Solution • Because there is always more than one possible solution to a problem, it is useful to compare and test designs	Tested different materials and compared how they stood up to a model of rainfall Discussed the different roofs that were tested and identified the best-performing roofs and the properties they had in common
Crosscutting Concept	
Structure and Function	Examined how the properties of different materials affect how they function, while testing properties under rainfall Discussed why a rigid roof is important to hold up against rainfall (can be extended to snowfall) Compared materials and explained what made them suited or not suited for use as a roof material

Source: NGSS Lead States 2013.

Acknowledgment

An earlier version of this activity was developed, in part, for the Smithsonian Science Education Center's STC kindergarten curriculum.

References

Biological Sciences Curriculum Study (BSCS) and International Business Machines (IBM). 1989. *New designs for elementary science and health: A cooperative project between Biological Sciences Curriculum Study (BSCS) and International Business Machines (IBM).* Dubuque, IA: Kendall Hunt.

Lachapelle, C. P., C. M. Cunningham, J. Facchiano, C. Sanderson, K. Sargianis, and C. Slater. 2012. Limestone or wax? *Science and Children* 50 (4): 54–61.

National Research Council (NRC). 2012. *A framework for K–12 science education: Practices, crosscutting concepts, and core ideas.* Washington, DC: National Academies Press.

NGSS Lead States. 2013. *Next Generation Science Standards: For states, by states.* Washington, DC: National Academies Press. *www.nextgenscience.org/next-generation-science-standards.*

Internet Resource

Characteristics of Materials

www.bbc.co.uk/schools/scienceclips/ages/7_8/characteristics_materials_fs.shtml

---

NSTA Connection

For a homework rubric and worksheet, as well as blank POE worksheet, visit *www.nsta.org/SC1601.*

Measuring Success

Second Graders Design, Build, Test, and Improve Tools to Map a Waterway

By Tori Zissman

Field trips are wonderful opportunities to expand student learning, but the bus rides can be challenging. Perched in the first row, teachers attempt to guide the driver while tossing repeated reminders of safe bus behavior to the students in back, inevitably arriving at the destination flustered and possibly nauseated. My colleagues and I addressed the issue of stressful bus rides by focusing our second-grade studies closer to home, within walking distance of school. We developed a series of units integrating science, social studies, math, and literacy around the single focus of our local river. Our first order of business: Find the river!

We begin our yearly river unit by collecting questions students have about the river. Some of the things students would like to know typically include the following: Where does the river start? Where does it end? Where does it come closest to our school? These questions launch a mapping unit that culminates in our first engineering design challenge: create a tool to measure length, width, or depth of the river. As students design, construct, test, and improve their tools, they build a deep understanding of

the challenges and rewards of the engineering process. They learn how to collect, analyze, and evaluate data and gain a greater understanding of the issues encountered while trying to create a map or model of a particular location. This design challenge engages students in conversations about what it means to be "measurable" and encourages them to examine what is "measurable" about a river. This guided-inquiry activity also promotes collaboration and sets the tone for a year of hands-on exploration with real-world applications.

A five-step design process provides the structure for our activities. Based on an engineering curriculum developed by the Museum of Science in Boston called Engineering is Elementary (see Internet Resources, p. 156), the activities guide students to ask, imagine, plan, create, and improve their measuring tool designs over a series of four lessons (EiE 2013; see Figure 19.1, p. 150). The lessons address engineering design performance expectations noted in the *Next Generation Science Standards* (*NGSS*). K-2-ETS1-1 explicitly notes that students who demonstrate understanding can "ask questions, make observations, and gather information about a situation people want to change to define a simple problem that can be solved through development of a new or improved object or tool" (NGSS Lead States 2013). These expectations are carried through succeeding levels of the standards, gaining specificity and depth for older students. Applying engineering performance expectations to the close study of the structure of a river also demonstrates the crosscutting concept of the interdependence of science, engineering, and technology in the *NGSS* (NGSS Lead States 2013). This chapter describes our design activities in detail as they evolved over a period of five years. It provides a model for teachers in the elementary grades who wish to replicate the process in their own settings.

FIGURE 19.1

Engineering Design Process

Engineering is Elementary
www.mos.org/eie
© Museum of Science, Boston

Setting the Stage

To prepare for the measuring tool design challenge, students usually needed to review map concepts and skills. With the rise in the use of global positioning system tools and internet resources, student exposure to paper maps is increasingly limited. In the days leading up to the design challenge, we engaged students in lessons to activate background knowledge, revisit concepts learned in earlier grades, and introduce new vocabulary. Students interacted with a variety of maps of our home state. They compared different types of maps, such as overhead and topographic maps. Students reviewed map conventions through a read-aloud of *Mapping Penny's World* by Loreen Leedy (2000), and they became cartographers themselves, working with partners to draw overhead maps of our classroom.

When interactive technology was introduced to our classrooms, we added a new activity to the unit. Using the whiteboard, students visited the website of the Norman B. Leventhal Map Center at the Boston Public Library (see Internet Resources, p. 156). Through the website, we were able to access a bird's eye–view map of our river created in 1900. Students commented on the river's twisting shape and learned a new word—*meander*. The website has a "zoomify" feature that allows students to take a closer look at small sections of the map. Students were especially excited to try to find our school. They discussed the cartographer's choices about what to include and what to leave out. They were puzzled by how the map was made, given that airplanes and Google Earth had not yet been invented.

Lesson #1: Engage

Once we had completed our review activities, the time was ripe to introduce the engineering design challenge to the students. The phases of the design challenge fit readily into the BSCS 5E Instructional Model of engage, explore, explain, elaborate, and evaluate (Bybee et al. 2006). We engaged students by informing them that we had been asked to map a section of the river near the school. To create the map, we would need to measure the dimensions of the river at the site. We let students know that the measuring work would be done in three small groups, and we introduced new vocabulary, including the words *length*, *width*, and *depth* to describe a few of the measurable dimensions of the river. One group would measure the length of the river between two easily recognized landmarks. Another would measure the depth of the river, and the third group would measure the river's width. At this point, a number of familiar measuring tools were brought out for students to consider. The advantages and disadvantages of using a ruler, a meterstick, or a measuring tape to gather data were discussed. Students quickly realized that none of our tools really fit the task and that we would need to design new measuring tools. Teachers explained that an adult would guide each group through a five-step design process that would include asking relevant questions about the task and the site, imagining a number of different possibilities, and selecting the best options to build and test at the river. Students were also told that they would need to use readily available materials to create their tools.

As an exit ticket from the lesson, students were asked to rank their task preferences. Over the years, teachers engaged in guiding these activities found that motivation and learning increased when students were given some control over their participation. Whereas one student might relish the possibility of getting her hands wet while measuring the width of the river, another might prefer staying on dry land to measure the length of the river along the bank. Teachers worked with these preferences to create three small groups of six to eight students, representing a mix of abilities. Extra adults were recruited from learning-support staff to allow for one adult facilitator to work with each group. Because this project fell early in the school year

and required collaboration, we found it especially helpful to involve staff members who usually assist with behavioral and social issues. All staff members working with the students were briefed about expectations for the project.

Lesson #2: Explore

In the second session, students delved deeper into the engineering design process (EDP). They continued to ask questions while they extended their thinking by brainstorming ideas (imagine) and choosing one design to develop in detail (plan). In this exciting phase, students rolled up their sleeves and got to work. Adult facilitators explained that each group was required to turn in four documents by the end of the lesson: a written description of the tool it planned to build, a drawing or diagram of the tool in use, a step-by-step explanation for how to use the tool, and a materials list (see NSTA Connection, p. 156). Figures 19.2 and 19.3 (p. 152) show one group's completed materials list and step-by-step explanation, respectively. Although a whole group can complete the worksheets together, we found it helpful to break into working teams. Specific jobs such as writer, artist, and supplier could be assigned by readiness, or students could volunteer based on interest.

As each group focused on its specific measurement task, additional questions emerged. A group measuring the river's width brainstormed ways to stretch a rope from one riverbank to the other. One especially creative proposal was to use a baking soda and vinegar volcano to shoot one end of the rope across! Another group wondered if they could use a boat to reach the middle of the river to measure depth. This group discussed whether the measurement had to be taken from the middle of the river at all. Would a measurement from near the riverbank yield the same results? Teachers and adult facilitators

19

FIGURE 19.2

List of One Group's Materials

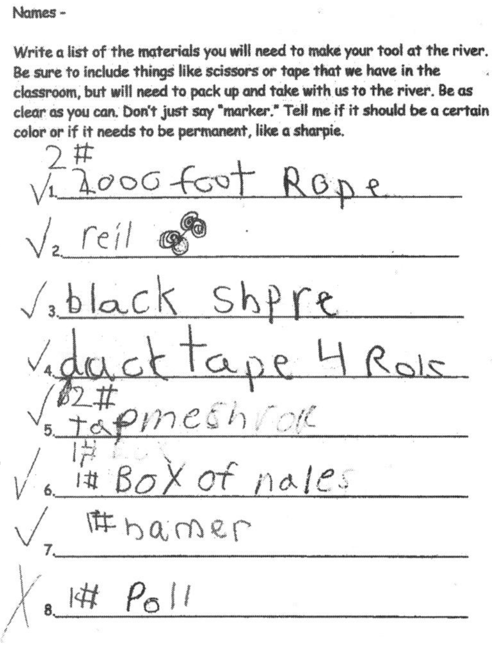

Names -

Write a list of the materials you will need to make your tool at the river. Be sure to include things like scissors or tape that we have in the classroom, but will need to pack up and take with us to the river. Be as clear as you can. Don't just say "marker." Tell me if it should be a certain color or if it needs to be permanent, like a sharpie.

1. 2000 foot Rope   2#
2. reil
3. black shpre
4. duck tape 4 Rols
5. tapmeshrok   2#
6. Box of nales   1#
7. hamer   #
8. Poll   1#

FIGURE 19.3

Step-by-Step Explanation for Tool Use

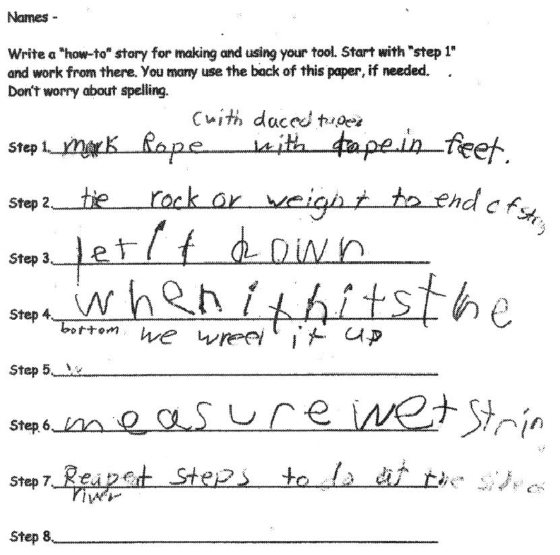

Names -

Write a "how-to" story for making and using your tool. Start with "step 1" and work from there. You many use the back of this paper, if needed. Don't worry about spelling.

Step 1. mark Rope with tape in feet. (with duced tape)

Step 2. tie rock or weight to end of string

Step 3. let it down

Step 4. when it hits the bottom we wreel it up

Step 5.

Step 6. measure wet strin

Step 7. Repeat steps to do at the sides river

Step 8.

Lesson #3: Investigate and Explain

guided students to additional resources as questions arose. For example, careful observation of photos of the site revealed a bridge that students could use to access both the center of the river and its far bank. Groups happily incorporated this new information into their plans.

These activities correlate with disciplinary core idea ETS1.B: Developing Possible Solutions, which is found in the K–2 engineering design strand of the *NGSS* (NGSS Lead States 2013). The text notes that "Designs can be conveyed through sketches, drawings, or physical models. These representations are useful in communicating ideas for a problem's solutions to other people" (NGSS Lead States 2013). In addition, review of these planning documents at the end of the lesson provided important assessment data, allowing classroom teachers to monitor small-group progress and to redirect any students who were off target or off task.

Once teachers had gathered the materials each group requested, it was time to visit the site to build and test the tool designs. Before leaving, we discussed behavioral expectations and safety rules, stressing that students were not to enter the water. We made sure students were able to identify poison ivy, which we knew was present at the site. Students had also been encouraged to wear long pants on the day of the field trip to limit exposure. Because the river was close to school, a teacher was able to do a safety check of our proposed work area, selecting sections of the river that were safe for student use. Hand sanitizer was provided on-site, and students washed their hands thoroughly with warm water and soap upon our return to school. Follow your school guidelines for field trips and see NSTA Connection (p. 156) for further tips about field trip safety. Because of the hands-on nature of the project and the age of our students, we arranged extra adult help from parents for the trip.

On-site, each group reviewed its planning documents and began building its measuring tool. At the river, the particular challenges of working with water were more apparent. Exploring the river's width, one group measured regular intervals along a rope using markers from the classroom. When the rope got wet, the marker washed off. They ultimately remedied the problem by tying knots in the rope instead. Another team measuring depth struggled because the current was moving their rope. The loose end of the rope kept floating downstream. A teacher recorded the students' discussion of the problem.

Student 1: Look! It's going diagonal. Too much string is going in. We need a buoy.

Student 2: Maybe something is blocking the bottom?

Student 3: Maybe it hit a fish!

Ultimately, they searched the river bank for heavy objects and tried tying different items to the rope until the end finally sank. In their discussion, members of this group were able to describe their observations, identify and explain issues, and develop new solutions. Students in every group continued to ask, imagine, and plan, but now they extended their learning by working together to create and improve their tools. These types of activities align with the *NGSS* science and engineering practice Planning and Carrying Out Investigations, which specifically encourages students to "Plan and conduct an investigation collaboratively to produce data to serve as the basis for evidence to answer a question" and falls under performance expectation 2-PS1-2: "Analyze data obtained from testing different materials to determine which materials have the properties that are best suited for an intended purpose" (NGSS Lead States 2013).

At this stage, adult facilitators introduced a new piece of the process. Each group needed to consider how to record the testing and measurement data it was collecting. Although the primary focus of this activity is the EDP, data collection is an important component of the testing phase. Testing records served as a road map for group work. They were also useful as an assessment tool for classroom teachers monitoring student use of the five-step EDP. Groups were encouraged to answer key questions: How well did the tool work? What changes were made to the tool? Did the changes solve the problem? One group tried two different methods to measure length, arriving at very different results. They realized the discrepancy was an important clue to the effectiveness of their design and further work was needed.

Lesson #4: Elaborate and Evaluate

Back at school, teams reflected on their experiences at the river. One group that also had a problem using marks on their rope made adjustments to their diagram to reflect their choice to use tape instead, a decision made at the river (see Figure 19.4, p. 154). Some groups used this time to stretch out their measurement tools in more familiar locations to get a sense of comparative size (relative to the playground climber, for example). One year, a group measuring the river's width found that the rope they had used stretched halfway across the parking lot. This was much larger than they had expected. Because it was difficult to measure the rope safely in that setting, the students decided to fold the rope evenly four times. This gave a length of rope that fit on the school's front porch. Students measured this length in standard units and then added that number four times to reach their final figure. As part of the reflection process, each team shared its measurement data and experiences with the class.

FIGURE 19.4

Diagram of the Tool

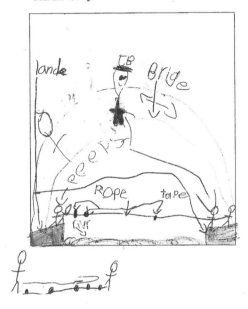

Names

Draw a diagram or picture of the tool in action. Be sure to label the parts of the tool and anything in the picture that will help me understand how your tool works. For example, if you drop your tool off the bridge, you should label the bridge.

Students were encouraged to ask questions and make suggestions.

Writing prompts or science notebooks may be easily incorporated at this stage. In our case, students were given individual project report forms to complete (see NSTA Connection, p. 156). Coupled with observation notes, these reflections formed the basis for student assessment relative to this project. One question on the reflection form asked students to describe how they improved their tool at the river. One student wrote, "Instead of tying the rock to the rope, tape it and tie it. We also had to put it in the way the river was flowing." This student records two different changes made to the tool to improve its effectiveness. Another student answered the same question by writing, "We did tally marks instead of duct tape to save time." Not only does he describe the

change made, he also explains why the change was needed. Both students demonstrated a deep engagement with the design process.

Epilogue

We had introduced the design challenge by telling students we were going to make a map of one section of the river. Because two classrooms executed the design challenge at the same time, it was possible for students to compare their findings with another group measuring the same dimension of the river. Often, the data from the two groups did not match. Ultimately, the students concluded that our tools were not up to the task and we did not have enough accurate data to create our map. Students were surprised to find that sometimes engineers' earnest efforts do not lead to success on the first—or even the second —try. This was often the perfect point in the process to bring in a scientist or other expert to discuss his or her own work. One year, we were fortunate to have a classroom visit from a parent who built and launched satellites providing geographic data to companies around the world.

The true purpose of this activity was to introduce students to a five-step EDP. We would use this same process to address two additional design challenges during the school year. The project built excitement for science learning and got students talking and listening to one another. Students used their investigations to find answers to real-world questions in a collaborative setting. They gathered data, identified important information, and revised their explanations to gain new insight and knowledge. They also exercised math skills and developed a deeper understanding of the connection between the visual representation of a map and the actual site itself. Although the work took planning— and was messy at times—the positive results were many and well worth the investment.

Connecting to the *Next Generation Science Standards*

The materials, lessons, and activities outlined in this chapter are just one step toward reaching the performance expectations listed below. Additional supporting materials, lessons, and activities will be required.

2-PS1 Matter and Its Interactions *www.nextgenscience.org/2ps1-matter-interactions* **K-2-ETS1 Engineering Design** *www.nextgenscience.org/k-2ets1-engineering-design*	**Connections to Classroom Activity**
Performance Expectations	
K-2-ETS1-1: Ask questions, make observations, and gather information about a situation people want to change to define a simple problem that can be solved through the development of a new or improved object or tool	Gathered and researched information and planned and conducted an investigation concerning the measurement of a near-by river Concluded that a new type of tools is necessary for measuring in the water
2-PS1-1: Plan and conduct an investigation to describe and classify different kinds of materials by their observable properties	Asked questions, made observations, explored materials, and tested their viability for solving the problem
Science and Engineering Practices	
Developing and Using Models	Created sketches and drawings of the possible tools to be used for measuring
Planning and Carrying Out Investigations	Conducted an investigation to collect data using the measurement tools designed specifically for the river
Analyzing and Interpreting Data	Recorded measurable data concerning the tools and results, shared results with classmates, analyzed the data and identified the strategies and components that were most successful
Disciplinary Core Ideas	
ETS1.A: Defining and Delimiting Engineering Problems • Asking questions, making observations, and gathering information are helpful in thinking about problems.	Asked questions, made observations, explored materials, and tested their viability for solving the problem Investigated and considered appropriate materials to use in designing and building a measuring tool
PS1.A: Structure and Properties of Matter • Matter can be described and classified by its observable properties.	Investigated and identified the characteristics of materials that provide the attributes needed to create a measuring tool to be used in a river
Crosscutting Concept	
Structure and Function	Analyzed individual materials for structure and function properties, determined how those materials could be combined to create a tool for measuring the river

Source: NGSS Lead States 2013.

References

Bybee, R. W., J. A. Taylor, A. Gardner, P. Van Scotter, J. Carlson, A. Westbrook, and N. Landes. 2006. *The BSCS 5E Instructional Model: Origins and effectiveness.* Colorado Springs, CO: BSCS.

Engineering is Elementary (EiE). 2013. The engineering design process. Museum of Science, Boston, MA. *www.eie.org/content/engineering-design-process.*

Leedy, L. 2000. *Mapping Penny's world.* New York: Henry Holt.

NGSS Lead States. 2013. *Next Generation Science Standards: For states, by states.* Washington, DC: National Academies Press. *www.nextgenscience.org/next-generation-science-standards.*

Internet Resources

Engineering is Elementary
www.mos.org/eie

Norman B. Leventhal Map Center at the Boston Public Library
http://maps.bpl.org

NSTA Connection

See *www.nsta.org/SC1310* for the reflection and measuring tools worksheets and field trip safety tips.

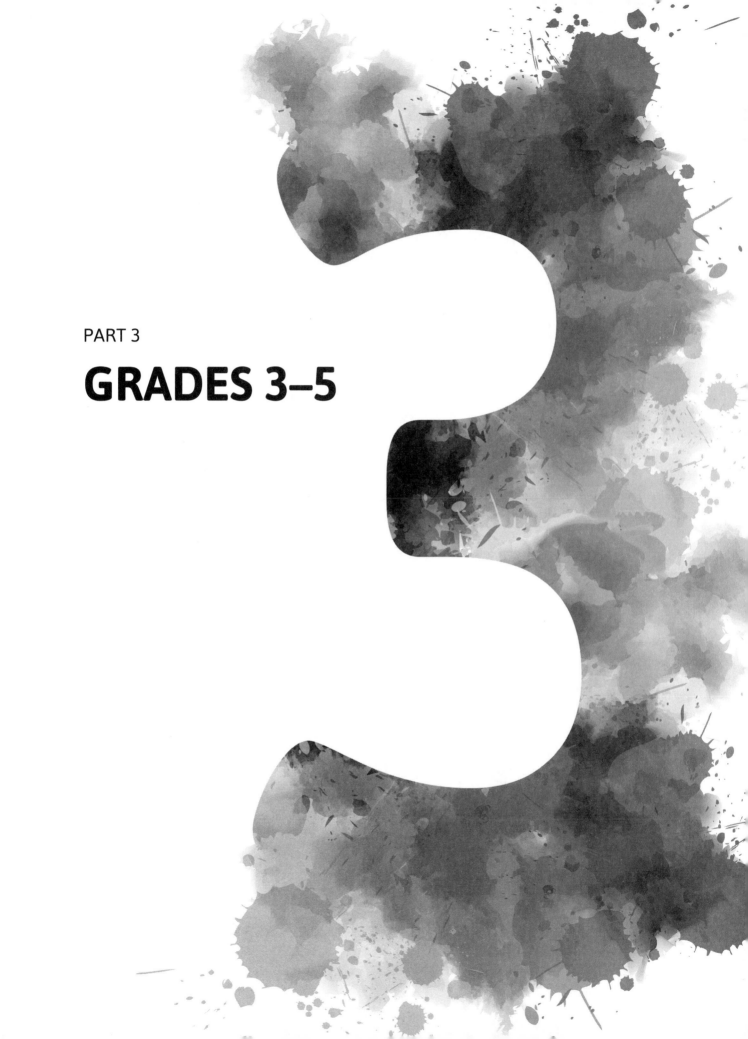

PART 3

GRADES 3–5

3 — 4 — 5

20

Think It, Design It, Build It, Test It, Refine It

A Unit on Water Quality Ends With a Water Filtration Engineering Design Activity

By Barbara Ehlers and Jeannie Coughlin

It was wild, crazy, and noisy, but it was a good noise—the noise of learning. That was our reaction when we co-taught a unit titled "Preserving and Protecting Our Water Resources." After learning about the amount of available freshwater and studying water quality problems and issues, our fourth graders were concerned about water pollution and curious about how to clean water. They dove into engineering by designing, building, and testing their own water filtration devices. Integrating engineering into the study of water quality was a natural finale for the unit. The students were able to apply what they had learned about water properties and water pollution into this engineering challenge.

Background

Through myriad hands-on activities, students explored the properties of water such as density, surface tension, and solubility, and its three forms—solid, liquid, and gas. Learning about the amount of water on Earth and the small amount of freshwater available for our use was eye-opening for these fourth graders as

evidenced by this student comment: "We might run out before I am 35 years old!"

To help illustrate the amount of Earth covered with water, students took part in a simple, engaging game in which they tossed an inflatable globe and recorded whether their left pinky was touching land or water when they caught the ball. Their findings mirrored the actual percentage of Earth's surface covered by water: 70%. Project Wet's water cycle game "The Incredible Journey" (see Internet Resources, p. 167) enhanced their knowledge of evaporation, transpiration, precipitation, and condensation.

Next, students researched present-day problems and issues with water quality and became very interested in learning how to preserve the precious water available. Students read *Riparia's River* (Caduto and Pastuchiv 2011) and newspaper articles about local water quality problems and issues in Iowa. (Teachers may wish to have students research and locate articles from their own geographic area.) They were able to integrate the information from the various texts to write and speak about the subject knowledgeably as indicated in the *Common Core State Standards* (*CCSS*), *English Language Arts* (CCSS.ELA-LITERACY.

R1.4.9; NGAC and CCSSO 2010). They recorded comparisons and analysis of the water quality problems and issues in a chart (see NSTA Connection, p. 167) and were able to draw evidence from literary or informational texts to support analysis, reflection, and research (CCSS. ELA-LITERACY.W.4.9; NGAC and CCSSO 2010). The completion of the table offered us an avenue of formative assessment to gauge our students' understanding of water quality problems and issues.

The personal correlation the students made with the readings was an outcome we did not expect. Through their readings, the students were able to obtain and combine information about ways individual communities use science ideas to protect Earth's resources and environment. News articles indicated water quality problems in Iowa and what was being done to address the problems. This prompted the students to start talking about cleaning water. In their research, they read the same amount of water exists that we have always had—it is used over and over, but a lot of the water is unusable for a variety of reasons. Student wanted to know how water is cleaned for reuse and were determined to find out more about how to clean water. Realizing this would be a perfect opportunity to incorporate an engineering design activity into our science curriculum, we began thinking of possible ways to allow for exploration and student involvement in the designing, constructing, and testing of water filtration systems. The classroom was abuzz with ideas and suggestions. "What should we use to build them?" "We have lots of plastic water bottles, let's use those."

Getting Started

An October 2013 *Science and Children* article entitled "Minding Design Missteps" (Crismond et al. 2013) provided a starting point for our plunge into engineering design. In this article, the authors point out several common errors typically made by beginning designers. Because our fourth graders had not previously experienced engineering design, we kept these missteps in mind as we structured this first engineering adventure. One misstep involved students beginning their designs before they understand the problem. To find out what students already knew about water filtration, we decided to use student ideas to construct the knowledge component (K) of the know, want to know, learned (KWL) model. Their initial ideas included "Water goes through different stages to get clean"; "A filter is like what some people have in their pool to help keep it clean"; and "Woodchips can separate water and pollution." Realizing the importance of understanding the problem, students filled out the W section of the KWL chart with what they wanted to know. Their questions included, "How well does it work?" "How can we tell how clean the water is?" and "Where does all the bacteria go?" (see NSTA Connection, p. 167, for the complete KWL chart). This formative assessment provided us with more information about their initial thoughts on filtration as well as their prior knowledge.

Another common mistake in engineering design that Crismond and his colleagues highlight is that students tend to begin tinkering with materials before they fully understand problem constraints and how they will measure whether their designs actually work. These factors, along with some of the questions students wondered about, had to be addressed before we could begin our design. We discussed the types of pollutants commonly found in water. Students referred to their research and realized that the polluted water they had read about looked and smelled bad. They suspected that filtering the water would improve the appearance and the odor. They also knew they had to be able to provide proof of the results, and odor would be difficult to indicate but appearance of the water would be easier. Some were familiar with filtering of water in their homes for drinking purposes and for pets. In other words, we defined the problem in more detail than merely saying we want to clean the water.

Think It

Students brainstormed how they might use plastic water bottles to design a water filtering system. Their plan was to design a solution for filtering polluted water and to test how their water filtration devices performed when polluted water was poured through them. They brought water bottles of various shapes and sizes. The students pondered how their devices might work. The "Think It" portion began with the students sharing their ideas and suggestions with their classmates about how to design and construct the water filtration devices. In terms of construction, they realized they needed materials inside the water bottles that the water would pass through but the dirt and other pollution particles would not. Some of their ideas included placing holes in the bottom of water bottles and using coffee filters, sponges, towels, and window screening.

Through these brainstorming sessions, it became obvious that students had some common understandings about what they would do. These became our design constraints. Students agreed that all filtration systems would adhere to the following constraints:

- Designs could use as many water bottles as the engineers wanted.

- The filtration materials must fit in the water bottle(s).

- Water must go in the top, be carried by gravity, and come out the bottom.

The most challenging question was how we could tell the water was clean. The next question that arose was about what would constitute success. Rather than just telling students how we would measure the cleanliness of the water, we let them decide. Students had many ideas, some of which included the following:

- We could tell by how the water looks (make a drawing before and after we filter, draw a cup-shape on our paper and color it the color the polluted water is both before and after filtering, or take pictures).

- We could check how clear it is in the beginning and at the end.

- We can look at what was caught in the filter.

- We could use the turbidity tube.

- We could smell it. How do scientists test smells?

- We could use pH strips to check for acids and bases and neutral.

Students remembered what they had learned about turbidity tubes when a guest speaker had demonstrated them to the class earlier in the unit study. Most of the students did not know about using pH strips to test water as suggested by a classmate, so we researched pH strips and explored how to use them. The students gathered sample liquids—water, dish soap, tomato juice, a kiwi-strawberry drink, and laundry detergent—to test if they were acidic, basic, or neutral and used a pH chart to check the results. Knowledge about pH and how the scale worked was definitely going to help in determining the success of the filters. Always use splash goggles when working with any liquids in the classroom.

A student asked, "What are we looking for when we are done with our water filtration systems as far as pH"? Another replied, "We hope we get in the ballpark of neutral." In the end, students agreed on the following measures of success:

- The closer the amount of water we get out to what we put in, the better the design.

- The closer the pH of the water is to 7, the better the design.

- The clearer the water is, the better the design.

They also decided on some safety rules everyone had to follow:

- Do not drink the water before or after filtration.

- Wear goggles when using your water filtration devices.

- Use extreme caution when cutting the water bottles—use scissors, not a knife. Only cut the water bottles under adult supervision.

- Wash your hands before and after the process.

Here is the list of materials the students identified and wanted available as they created their filters:

- Plastic water bottles
- Coffee filters
- Activated charcoal
- Cotton balls
- Macaroni
- Baking soda

- Calcium carbonate (One of the students suggested this because he thought they had used it at home to filter water.)

- Gravel

- Sand

- Window screen

- pH paper and indicator scale

- Measuring cup

- Masking tape

- Spoons

Design It

With constraints understood, measures of success established, and safety guidelines set out, our next step was to begin to design a filter system. First, each student drew a picture of the water filtration device he or she wanted to build, labeling it with the filtering materials. Next, students met with a partner designated by the teachers and negotiated what design they would actually use for their devices. As mentioned in the *NGSS* standard, communicating with peers about proposed solutions is an important part of the design process, and shared ideas can lead to improved designs (3-5 ETS1-1: Engineering Design and ETS1B: Developing Possible Solutions; NGSS Lead States 2013). This also aligned well with the Mathematical Practices from the Common Core Math Standards, to construct a viable argument and critique the reasoning of others. Again, the classroom was alive with discussion. At first, most students tried to convince their partner to use their design. After furthur discussion, they began to negotiate a mixture of the designs to produce one design based on what they thought might work best. Their personal drawings and the drawings that resulted from their negotiation, as well as the class discussion, served as another formative assessment to help us understand their thoughts about the filtering process and as a chance to ponder using another *CCSS Mathematics* practice—modeling with mathematics. We were quite impressed with the discussion that ensued about designing the water filters: "I think it would work better if we put the larger materials first, then the smaller ones"; "We should put coffee filters between all the layers of materials"; "Let's put the cotton balls first to take out more stuff"; "We can't put the baking soda on the bottom, it will come right out"; "Should the filtering materials be close together or should there be space between them?" "How can we fit the water bottles inside each other?" and "What if it leaks?"

Students' prior knowledge of filtering materials helped in their design process. Comments students made include the following:

- "I know that coffee filters allow water to pass through, but the coffee grounds stay in the filter. I think the coffee filters will take stuff out of the water"

- "I use a filter in my fish tank. It looks like cotton balls and I have to change it when it looks dirty. It really takes out a lot of gunky, gross stuff."

- "My mom puts baking soda in the refrigerator to absorb smells."

- "My mom puts baking soda down the sink to take out the smell down there."

- "On the bag of calcium carbonate, it says it cleans water in aquariums"

Students did not seem to know anything about activated charcoal, so they researched that. According to How Stuff Works (see Internet Resources, p. 167), "activated charcoal is charcoal that has been treated with oxygen to open up millions of tiny pores between the carbon atoms." This information helped students understand that activated charcoal removes some chemicals by attaching to the charcoal, but it does not remove everything. It also stops working when those pores are filled up. One of the most powerful learning opportunities was when students synergized by combining ideas with their partners. They were completely engaged, which led to better designs because of the thought they put into the design process!

Build It

After each pair agreed on one design and sketched the idea, they built their systems. However, what they thought would work on paper did not necessarily work when they built it. Some filtration systems did not resemble what they had drawn on paper. Sometimes it fell apart and they had to rebuild, giving them practice in the *CCSS Mathematics* practice of making sense of problems and persevering in solving them. Many students wished they had duct tape instead of masking tape because duct tape was waterproof. We discussed how scientists and engineers often make changes to their original plans. Students again made modifications of the system they had drawn. When they were finished building, students were very eager to try out their designs with the polluted water.

Test It

Students tested their designs using one cup of polluted water that contained soil, salt, dish soap or vinegar, and tap water. Some groups ran the water through their filtering systems several times. They colored a cup shape on the back of their KWL sheet the color of the polluted water and did the same when they were finished (see Figure 20.1, p. 164). They also checked the pH of the water before and after filtration, exhibiting their knowledge of using appropriate tools strategically, as mentioned in the *CCSS Mathematics* practices. As outlined in their success criteria, their goal was to make the water clear and neutral. Although they realized they might not achieve that, they definitely wanted to see improvement in the water quality. They also measured the water after the filtering to determine if they lost any water in the process. The excitement in their voices was evident as they proclaimed their results. Some talked about the number of times they ran the water through their filters. "We sent the water through 15 times and it was almost clear and had no smell"; "We filtered our water 10 times and it was quite clear"; "The first time the water changed a lot, but then it did not change. I think we needed more charcoal." Others described the change in pH. They also noted the amount of water they started and ended with, which connects to the *CCSS Mathematics* standard to solve problems involving measurement and represent and interpret data. Most important, students started to think about the why of their results. "We needed more layers of filtering materials. Ours went through quickly. The smell was disgusting. We started with 8 oz. and ended with 7.5 oz." Their next thoughts centered on modifying their devices, aligning with the *NGSS* Crosscutting Concept Cause and Effect, which states that cause and effect relationships are routinely identified, tested, and used to explain change. The drawings of the water before and after, the pH changes, the data denoting the change in amount of water, and the comments they wrote served as another mode of formative assessment to determine what they were learning in the process.

Refine It

After numerous trials of their water filtration systems, it was time for students to explain what they had done, what worked, and what did not.

FIGURE 20.1

Sketch of Polluted Water and Filtered Water

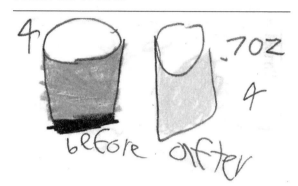

Students focused on the measures of success they had established as a class. What was the pH of the water at the beginning and at the end? How much water did we start with? How much did we end with? What did it look like? Students wrote the data for their measures of success on the back of their KWL chart. They were also asked to write some ideas for how to improve their designs. As we brainstormed ideas for refining designs, many additional ideas emerged. Students were bubbling with ideas about how to modify their water filters.

To help reach closure on our engineering design adventure, the students filled out the *L* portion of their KWL chart, explaining what they had learned in the process of designing, building, and using their water filtration systems. Students surprised us with the unsolicited connections they were able to make between what they did and learned during their design activity and some of the realities of real-life water filtering. As a summative assessment, the students answered the following questions:

1. What was the purpose of this design project?

2. How did you and your partner negotiate the design of your water filtration device?

3. What changes did you make to your original design when you actually built the water filtration device? Why?

4. How did your filter work?

5. What modifications did you make after trying your original design? Why?

6. What were your criteria for success?

7. How did you decide which filtering materials were most successful?

8. What materials did you wish you had that you did not?

9. What three conclusions can you draw from this design experience?

10. How could you apply what you learned in this lesson to real life?

Conclusion

From our point of view as teachers, this was a deep learning experience for the fourth graders.

Building an engineering design activity into our water study unit as a culminating experience was a win-win situation. Students were already very curious to learn about water filtration after their study of water properties and water quality problems and issues. They were concerned about what their future will be like if clean water is not readily available. Getting them actively involved in design and construction of a model that actually did change the polluted water helped them understand more about not only filtration but also engineering design. Every student was actively engaged in the learning. The excitement in the room was invigorating. Two students were writing down all of the things they used in their filtration system so they could gather them and try another design this summer. They reflected, "We could do this all day with the water in our creek! I am going to build a water filtration system at home this summer and try different ideas." What a fitting comment for our first dip into engineering design. We will both be on the lookout for ways to incorporate engineering into more of our science units in the future.

Connecting to the *Next Generation Science Standards*

The materials, lessons, and activities outlined in this chapter are just one step toward reaching the performance expectations listed below. Additional supporting materials, lessons, and activities will be required.

5-ESS3 Earth and Human Activity *www.nextgenscience.org/5ess3-earth-human-activity* **3-5 ETS1-1 Engineering Design** *www.nextgenscience.org/3-5ets1-engineering-design*	**Connections to Classroom Activity**
Performance Expectations	
5-ESS3-1: Obtain and combine information about ways individual communities use science ideas to protect the Earth's resources and environment	Researched and analyzed water quality problems and issues in their state Discovered possible solutions
3-5 ETS1-1: Define a simple design problem reflecting a need or want that includes specified criteria for success and constraints on materials, time, or cost	Designed and built water filtration devices Handled limitations of available materials and time
Science and Engineering Practice	
Asking Questions and Defining Problems	Wanted to learn how water can be cleaned Limited by materials and time Determined the criteria for success included pH level, appearance, and final volume of water
Disciplinary Core Ideas	
5-ESS3.C: Human Impacts on Earth Systems • Human activities in agriculture, industry, and everyday life have had major effects on the land, vegetation, streams, ocean, air, and even outer space. But individuals and communities are doing things to help protect Earth's resources and environments.	Investigated water quality problems and issues in their state and discovered that water quality problems are frequent and filtering is a method to improve water quality
3-5 ETS1.A: Defining and Delimiting Engineering Problems • Possible solutions to a problem are limited by available materials and resources (constraints). The success of a designed solution is determined by considering the desired features of a solution (criteria).	Limited to the list of available materials they had developed Determined success of filtration would be identified by pH level, volume, and appearance

5-ESS3 Earth and Human Activity www.nextgenscience.org/5ess3-earth-human-activity 3-5 ETS1-1 Engineering Design www.nextgenscience.org/3-5ets1-engineering-design	Connections to Classroom Activity
Crosscutting Concepts	
Systems and System Model	Drew a model of a water filtration device
Scale, Proportion, and Quantity	Measured water before and after filtering Checked the pH level before and after filtering Explained their findings about water filtration through measurement, pH, and color of the water after filtering
Influence of Science, Engineering, and Technology on Society and the Natural World	Acted like engineers by designing, building, testing, and revising their filtration devices

Source: NGSS Lead States 2013.

References

Caduto, M., and O. Pastuchiv. 2011. *Riparia's river.* Thomaston, ME: Tilbury House Publishers.

Crismond, D., L. Gellert, R. Cain, and S. Wright. 2013. Minding design missteps. *Science and Children* 51 (2): 80–85.

National Governors Association Center for Best Practices and Council of Chief State School Officers (NGAC and CCSSO). 2010. *Common core state standards.* Washington, DC: NGAC and CCSSO.

NGSS Lead States. 2013. *Next Generation Science Standards: For states, by states.* Washington, DC: National Academies Press. *www.nextgenscience.org/next-generation-science-standards.*

Internet Resources

How Stuff Works: What is activated charcoal and why is it used in filters?

http://science.howstuffworks.com/environmental/energy/question209.htm

Project Wet: The Incredible Journey

www.projectwet.org/resources/materials/discover-incredible-journey-water-through-water-cycle

Resources

Cobb, V., and J. Gorton. 2002. *I get wet.* New York: HarperCollins.

Kerley, B. 2006. *A cool drink of water.* Des Moines, IA: National Geographic Children's Books.

NSTA Connection

For the complete KWL chart, the water quality issues analysis chart, water filtration KWL worksheet, and the rubric, visit **www.nsta.org/SC1502.**

Community-Based Engineering

A Design Task Helps Students Identify and Find Solutions to a School-Yard Problem

By Tejaswini Dalvi and Kristen Wendell

If we take a minute and think, "How did I learn to drive or bake a cake?" we would all agree that the answer is "learning by doing," no matter what age we are. As adults, be it preparing for a particular profession or developing a new hobby, we learn by taking a plunge and doing it. For children, learning to use a spoon, ride a bike, or type is all about learning by doing. The experience of "doing" is involving, engaging, and exciting for children and allows for the "legitimate participation" that is the crux of meaningful learning (Lave and Wenger 1990).

As a team of science teacher educators working in collaboration with local elementary schools, we have been exploring opportunities for science and engineering "learning by doing" in the particular context of urban elementary school communities. Recognizing the unique assets and particular challenges that urban students and teachers face, we are developing a research-based community-based engineering (CBE) approach that can be incorporated into existing elementary science programs. CBE is an instructional strategy for elementary urban classrooms that provides a platform for children to participate actively in problem solving and construct new understandings and practices. It involves the finding and solving of engineering problems in students' local environment, such as their neighborhoods, community centers, or schools. For example, elementary students might notice a nearby vacant lot and design and build tools and techniques to prepare its soil for urban gardening. These engineering experiences engage students in responsibly framing a problem, planning a solution, and prototyping and testing artifacts. This approach also addresses the *Next Generation Science Standards' (NGSS)* call to integrate engineering and science learning and expose K–12 students in all sociocultural contexts in the United States to the practices and big ideas of engineering (NGSS Lead States 2013). The CBE experiences can connect to one or more units in the school science curriculum, reinforce disciplinary core ideas from the *NGSS*, and engage students in key science and engineering practices.

Module Overview

We have developed sample community-based engineering modules by pilot teaching with fourth and fifth graders at two different urban public schools. In collaboration with elementary science specialist teachers, we focused on two major problems within their school communities: watering plants when there is no easy water access and growing plants where there is very limited open space. Our team of teachers and researchers hoped these choices of community problems would help students understand science and engineering as a part of their community life and not as purely academic or abstract content. We also established formal learning objectives aligned to the following *NGSS* practices: (a) Asking Questions and Defining Problems and (b) Constructing Explanations and Designing Solutions. They also connect to the *NGSS* engineering design standards' performance expectations and disciplinary core ideas (NGSS Lead States 2013). We wanted students to improve their abilities in four areas:

1. *Unpack the problem.* Identify the problem, understand the need to solve it, and recognize constraints and criteria.

2. *Research and plan a solution.* Brainstorm ideas that could be potential solutions to the problem; use resources and discussions to plan a solution.

3. *Construct and test a prototype.* Test the planned solution by building prototypes or working models.

4. *Explain and redesign.* Explain and convey ideas to your peers; take feedback. Redesign the modeled solution according to testing results and feedback.

We conducted our module in four 90-minute sessions over a period of four consecutive weeks as follows:

- Session 1: Preview module and introduce engineering with a warm-up design challenge.

- Session 2: Unpack and explore the community-based problem.

- Session 3: Build and test solutions.

- Session 4: Explain and redesign solutions.

The CBE Module: Engineering a Water Transport System

Session 1
Teacher preparation: Identify potential community problems

We prepared for the first session by joining the school garden club teacher for a walk through the school yard. We asked her questions about the garden and took notes about the problems we saw. Together, we decided to focus the students' efforts on the problem of watering the upper-level garden beds. Because the upper beds were behind a fence, up a staircase, and across the playground from the spigot, they had no easy access to water. The teacher was excited about the prospect of students developing a system to solve this problem. We thought it offered many opportunities for scientific reasoning and for students to engineer working models—an opportunity to really apply science to construct things that actually function.

Introduce engineering design with a warm-up challenge

We began our first session by announcing to students that they would be working together to solve a problem in their school garden and by briefly describing the different phases of the project. We explained that the purpose of this activity was not only to help them learn to use engineering to solve a problem but also to have them work together to help their school. Then, we facilitated an informal whole-class discussion centered on the question, "What does it mean to do engineering?" We used this as a brief introduction to engineering and a platform to address students' notions that engineering refers only to products that require electricity, computers, or new technology and that engineers are people who "fix" and "build" things. By the end of this discussion we established with the students a working definition of engineering design: a process of creating solutions to human problems through creativity and the use of our math and science knowledge.

Students then completed a warm-up engineering challenge to get ready to solve a big problem out in the school garden. Students listened as we told a story based loosely on the children's book *Muncha! Muncha! Muncha!* (Fleming and Karas 2002): a gardener named Mr. McGreeley planted a vegetable garden, but every morning when he went to pick vegetables, he found half-eaten vegetables and destroyed plants. He saw animal tracks and realized that bunnies and squirrels were getting into his garden. We asked the students, "Did Mr. McGreeley have a problem? What was his problem? Could we use engineering to help Mr. McGreeley?"

In small groups, the students brainstormed ideas to solve the problem. They made drawings to plan a protection system for the vegetable garden. Each group then presented their design to the whole class. Here we introduced the students to the "Plus/Delta/Question" (+/Δ/?) method for offering peer feedback. This is a modification of the Plus/Delta feedback classroom assessment technique that helps identify what is going well in the class and what needs to be changed. It helps students think about what they should continue doing to learn well (PLUS) and what needs to be improved or changed (DELTA) (Helminski and Koberna 1995). After each design

presentation, each student in the class jotted down three things on a sticky note: a "+" telling one good thing about that design, a "Δ" suggesting one change to improve the design, and a "?" noting something unclear about the design. The students gave their sticky notes to the team presenting their design. The students used the "+/Δ/?" peer feedback to improve their designs and then built models of their garden protection systems with connectable building blocks. We purposefully restricted the materials to be used to introduce the factor of constraints as a part of the engineering design process to the students.

Finally, with the whole class we revisited the question, "What is engineering?" in the context of the activity they did. The students identified the engineering steps they took to help Mr. McGreeley as identifying the problem, thinking about different ideas to solve the problem, drawing plans, proposing designs, exchanging ideas, improving the design, and building and testing a prototype.

Our first session concluded with a closer look at the students' own garden out in the school playground. While the students walked through their school yard, they made a list of different gardening problems they observed. Any time the students visited the playground, we made sure they were fully supervised and we reinforced the playground safety rules before every session. See NSTA Connection (p. 177) for more safety tips for outdoor exploration.

Session 2
Unpack the community problem with students

The aim of our second session was to help students learn how to unpack an engineering problem. It included defining a central engineering problem and defining criteria for their solutions. We began with a discussion about the various problems the students observed in their school garden and those they thought might be solved by engineering design. Examples include the following:

- "People step on garden beds."

- "The beds need more flowers."

- "It gets dirty around the garbage area."

- "Balls go in the garden beds."

- "There is no water for the upper garden beds."

- "There is no dirt in the upper garden beds."

- "People pick flowers from the front beds."

- "Wood chips get into the garden beds."

We displayed each problem in large print on an index card at the front of the class. The students then sorted the problems in various groups, such as problems related to protecting plants or problems related to providing light. We then discussed the importance of choosing one problem that needed immediate attention and their collective efforts. Together we chose the watering problem as our focus problem. The students thought about the nature of this specific problem by considering the following questions (sample student responses are in quotation marks):

- Why do you think this garden problem is important to solve? "Otherwise the flowerbeds are just going to stand there. If there is anything planted we can't water it during vacations because we (students) aren't there to take turns and carry water."

- How will we know if we successfully solve the problem? "Whatever we plant will be growing well in the beds."

- What kinds of tests will the solution have to pass? "It has to save us time and all the effort of carrying the water."

- Are there any limitations on what we can design? "The amount of time we have to do it."

- What information or materials might we need to solve this problem? Or is there anything we need to learn before we solve it? "First, we will have to think what to construct, and then we can make a list of materials."

Brainstorm ideas

With the central problem now established, the students brainstormed to suggest possible solutions to the problem and made their first plans. In small groups, they made drawings and wrote descriptions. Each group shared their ideas with the class, solicited feedback, and documented their initial brainstorms on index cards.

Gather and document data

Our next step was to prepare the students to design a system with the potential to be actually installed in the school yard. After agreeing they needed more information about the size and shape of the school yard, the students went outside to take measurements and make maps using small whiteboards and dry-erase markers.

After measuring and mapping the outdoor space, the students created design diagrams for their watering systems and made a list of materials they would need to build their prototypes. We used the mapping activity to make students aware of the safety of other students when on the playground and encouraged them to consider this as design criteria for safety issues. Over the course of the next week, we gathered the materials listed by the students. We provided an alternative if a requested material was unsafe or too expensive and explained the reason for the swap to the students. We also added a few extra materials to support students' chance for success. Available to the students were scissors, cardboard, papers, duct tape, twine, rope, pulleys of different sizes, buckets, hand-operated water pumps, rubber sheets, empty sewing bobbins, plastic sheets, cartons, tissue roll tubes, and plastic containers.

We note that this is a point in the module where other accommodations can be made to support students with special needs. For example, the teacher can provide building instructions for an initial, minimally functional solution to the problem and ask the students to construct and improve it. During the design process, the materials and tools can be made simple or complex depending on the students' skill level. As the students complete their design, they can be asked to document their work through alternative means of expression, such as an interview with a teacher, a video, an audio recording, or a tablet computer presentation.

Connecting to the *Common Core State Standards*

English Language Arts
Speaking and Listening (Grade 4)

Comprehension and Collaboration

- CCSS.ELA-LITERACY.SL.4.1: Engage effectively in a range of collaborative discussions (one-on-one, in groups, and teacher-led) with diverse partners on grade 4 topics and texts, building on others' ideas and expressing their own clearly.

Presentation of Knowledge and Ideas

- CCSS.ELA-LITERACY.SL.4.4: Report on a topic or text, tell a story, or recount an experience in an organized manner, using appropriate facts and relevant, descriptive details to support main ideas or themes; speak clearly at an understandable pace.

Mathematics
Measurement and Data
Solve problems involving measurement and conversion of measurements. (Grade 4)

- CCSS.MATH.CONTENT.4.MD.A.1: Know relative sizes of measurement units within one system of units including km, m, cm; kg, g; lb, oz.; l, ml; hr, min, sec. Within a single system of measurement, express measurements in a larger unit in terms of a smaller unit. Record measurement equivalents in a two-column table.

Convert like measurement units within a given measurement system. (Grade 5)

- CCSS.MATH.CONTENT.5.MD.A.1: Convert among different-sized standard measurement units within a given measurement system (e.g., convert 5 cm to 0.05 m), and use these conversions in solving multi-step, real world problems.

Geometric measurement: understand concepts of volume. (Grade 5)

- CCSS.MATH.CONTENT.5.MD.C.3: Recognize volume as an attribute of solid figures and understand concepts of volume measurement.

Source: NGAC and CCSSO 2010.

Session 3
Fixing the problem: Execute the plans

The next session began with a brief Design Squad video clip (see Internet Resource, p. 177) showing high school students engineering solutions to another community problem. After watching the video, our students spontaneously referred to the engineering design practices that we had discussed earlier and pointed those out in the context of the video. They also compared and contrasted their own design process to the Design Squad participants' process, which although similar in structure, was in the context of an engineering problem faced by a local farm rather than a school garden. The students' thoughtful analysis of the Design Squad clip convinced us that the video had inspired them and provided an opportunity to see the engineering design process in action. The students also observed the use of various tools and safety equipment (including safety goggles) in the video. Our students' designs did not require the use of any potentially dangerous tools or materials and hence no specialized safety equipment was required. However we recognized this as a possible place to model safety procedures relevant to the chosen community problem, tools, or materials. We introduced and spoke to them about safety equipment, such as safety goggles for eye protection.

Next, each group shared their design for the water system with the whole class. As teachers, we identified key underlying science concepts for each design and asked the students to consider how they could apply their science knowledge and thinking to their systems. For example, we talked about pulleys and their role as simple machines to provide support to one of the groups who had pulleys included in their design of the watering system. The students then went back to their small groups to build prototypes with the materials we provided. After telling students which materials were alternatives to those on their lists (because of cost or safety factor), we explained that engineers need to take into account available resources while designing a solution, and limitations on resources are called "constraints."

One student group constructed a pulley and pump system to lift buckets of water to the upper garden beds and transfer the water easily through plastic tubing. Their system was designed to reduce the effort needed to carry heavy loads of water up the stairs. The second group created a miniature model of a conveyer belt system to transport containers of water safely across the school yard to the garden beds.

Session 4
Discuss and revise our models

At the beginning of our culminating session, each of the groups shared their prototype, and the whole class discussed the pros and cons of the designs. After this feedback session, the students refined their prototypes and documented all their revisions.

To wrap up, we displayed all the initial index cards (first suggested solutions) and had students compare those with their current prototypes. We reflected on the evolution of their designs and the nature of their engineering design process.

Evidence of Multifaceted Learning

While the students focused on solving their school garden problem, we used direct observation and collections of student work to formatively assess students' progress in three areas: engineering design practices, exploration and application of science and math concepts, and assuming the role of community problem solvers.

We assessed student growth in engineering design practices by reviewing students' design plans (Did they create multiple versions?), their feedback to each other (Did they write and talk about both strengths and suggestions?), their prototypes (Did they make improvements as they test and receive feedback?), and their documentation of final solutions (Did they create labeled diagrams that another engineer could follow?). See NSTA Connection (p. 177) for a rubric that can be used with this module.

We also asked students to reflect explicitly on what they learned about engineering. A response that showed strong evidence of student learning was, "When you engineer, you have to make a diagram, figure out what materials you need, build it to try it out, and solve its problems."

Strategies for assessing mathematics and science connections in a CBE module depend on the nature of the chosen community problem. In our problem, there were related science core ideas in the physical science and life science domains and important math concepts in measurement. As the students worked through the module, they did the following:

- Learned about pulleys as simple machines and explored double pulley systems to improve the usability of their design

- Discussed the watering needs of plants to understand the quantity of water and frequency of watering

- Explored concepts of weight and volume as they asked, "How much water do we need to carry?" and "How can we carry it?"

- Determined appropriate units of length as they measured distances across the school playground, made maps of the outdoor space, and used the measurements and maps to make informed design decisions

Finally, we observed that the students increasingly took on the role of problem solvers within their community. They discovered the power of both being collaborators and critical friends. One student said, "When we give each other feedback, it really makes our projects work better." They also commented on the importance of shared persistence: "Teachers should know that they shouldn't let kids give up; they have to help them keep on trying, keep on trying." Students also noticed and discussed other school yard problems and invited friends from other grade levels and collectively worked together. Students initially thought that all of these problems required a solution to be engineered, but later concluded that some problems could be solved by mere talk and creating awareness among other students.

Acknowledgments

This research was supported by the National Science Foundation under grant DRL-1253344. Any opinions, findings, and conclusions expressed in this material are those of the authors and do not necessarily reflect the views of the National Science Foundation. We are grateful to elementary science specialist Theresa Lee for her contributions to this learning experience, and we also wish to thank Anne DeVita and her students at Charles Sumner Elementary School for their collaboration on this work.

Connecting to the *Next Generation Science Standards*

The materials, lessons, and activities outlined in this chapter are just one step toward reaching the performance expectations listed below. Additional supporting materials, lessons, and activities will be required.

3-5 ETS Engineering Design *www.nextgenscience.org/3-5ets1-engineering-design* **5-LS1 From Molecules to Organisms: Structures and Processes** *www.nextgenscience.org/5ls1-molecules-organisms-structures-processes*	**Connections to Classroom Activity**
Performance Expectations	
3-5-ETS1-1: Define a simple design problem reflecting a need or a want that includes specified criteria for success and constraints on materials, time, or cost	Generated criteria for successfully solving the problem, and determined limits on the materials and time
3-5-ETS1-2: Generate and compare multiple possible solutions to a problem based on how well each is likely to meet the criteria and constraints of the problem	Constructed a prototype solution, explained the solution, and redesigned it according to results and feedback
Science and Engineering Practices	
Asking Questions and Defining Problems	Identified an engineering problem in the school yard
Constructing Explanations and Designing Solution	Researched and planned a solution to the problem in the community
Disciplinary Core Ideas	
ETS1.A: Defining and Delimiting Engineering Problems • Possible solutions to a problem are limited by available materials and resources (constraints). The success of a designed solution is determined by considering the desired features of a solution (criteria).	Considered problems, selected one to solve, identified constraints, posed solutions, and constructed and iterated on prototype solutions
ETS1.B: Developing Possible Solutions • At whatever stage, communicating with peers about proposed solutions is an important part of the design process, and shared ideas can lead to improved designs.	Explained their prototypes and shared ideas with classmates
LS1.C: Organization for Matter and Energy Flow in Organisms • Plants acquire their material for growth chiefly from air and water.	Discussed and analyzed the watering needs of plants to support growth
Crosscutting Concept	
Influence of Science, Engineering, and Technology on Society and the Natural World	Designed and built a watering system to meet the needs of the school community

Source: NGSS Lead States 2013.

References

Fleming, C., and B. Karas. 2002. *Muncha! Muncha! Muncha!* New York: Atheneum Books.

Helminski, L., and S. Koberna. 1995. Total quality in instruction: A system approach. In *Academic initiatives in total quality for higher education*, ed. H.V. Roberts, 309–362. Milwaukee, WI: ASQC Quality Press.

Lave, J., and E. Wenger. 1990. *Situated learning: Legitimate peripheral participation*. Cambridge, MA: University of Cambridge Press.

National Governors Association Center for Best Practices and Council of Chief State School Officers (NGAC and CCSSO). 2010. *Common core state standards*. Washington, DC: NGAC and CCSSO.

NGSS Lead States. 2013. *Next Generation Science Standards: For states, by states*. Washington, DC: National Academies Press. *www.nextgenscience.org/next-generation-science-standards*.

Internet Resource

PBS Design Squad Video Clip
http://pbskids.org/designsquad/video/green-machines

NSTA Connection

Visit *www.nsta.org/SC1509* for outdoor safety tips and the rubric.

Engineer It, Learn It

Science and Engineering Practices in Action

By Cathy P. Lachapelle, Kristin Sargianis, and Christine M. Cunningham

Children in America today spend most of their time in human-built spaces: home and school, playgrounds and city streets, public parks, and even farms. Children also spend most of their time interacting with technologies, from the banal (pencils and desks) to the new and flashy (iPads and cell phones). To succeed as citizens in the modern world, children need to know how such things come to be. When they understand that engineers design technologies—and when they understand the wide range of challenges that engineers address—children open their minds to new career possibilities.

Engineering is prominently included in the *Next Generation Science Standards* (*NGSS*; NGSS Lead States 2013), as it was in *A Framework for K–12 Science Education* (*Framework*; NRC 2012). The National Research Council (NRC) writes, "Engineering and technology are featured alongside the natural sciences (physical sciences, life sciences, and Earth and space sciences) for two critical reasons: (1) to reflect the importance of understanding the human-built world and (2) to recognize the value of better integrating the teaching and learning of science, engineering, and technology" (NRC 2012, p. 2).

For nine years, our team has been developing and testing engineering curricula for elementary students. We're engaged in this endeavor because, like the NRC, we recognize the importance of technological literacy. In this chapter, we'll show how science and engineering practices can be integrated into the elementary classroom by providing snapshots of activities from one of our STEM curriculum units. The unit, focused on aerospace engineering, challenges students to design parachutes for a spacecraft that will land on a planet with an atmosphere thinner than Earth's.

Science and Engineering Practices

The *NGSS* specifies that children should engage in eight science and engineering practices. In the following sections, we present scenes from an elementary school classroom that show children engaging in all eight engineering and science practices while they work on engineering parachutes. The activities we describe have been done successfully with students who are both experienced and not experienced with these practices. We also present examples in which these children, as they are engaged in engineering design, apply their scientific and mathematical knowledge and skills. We do this to illustrate how science, technology, engineering, and mathematics can be integrated, as the *NGSS* advocates.

Asking Questions and Defining Problems

The *Framework* notes that science and engineering have different goals. The goal of science is to create theories that explain how the world works, so scientists begin with questions related to this topic. The goal of engineering is to find a solution to a need or want, so engineers begin with questions that define the problem, describe what success will look like, and identify constraints on how the problem can be solved (NRC 2012, p. 56). The *NGSS* states that elementary students are expected to ask both kinds of questions (disciplinary core idea ETS1.A: Defining and Delimiting Engineering Problems; NGSS Lead States 2013).

In the Designing Parachutes unit, students ask both science and engineering questions. They ask about scientific phenomena related to parachutes, and they ask about the criteria for (and constraints on) their parachute designs (performance expectation 3-5-ETS1-1: "Define a simple design problem reflecting a need or a want that includes specified criteria for success and constraints on materials, time, or cost"; NGSS Lead States 2013). Now, let's go into the classroom and see what actually happens.

Mrs. A asks her third-grade students, "What information do you need to make sure that your team's parachute design is 'mission-ready'?" The students ponder this question. One girl responds, "Where is the parachute going? What planet?" Another girl jumps in, asking, "What can we use to build it?" A boy asks, "What is going to be attached to it? What are we dropping?" A girl adds, "How do parachutes slow you down, like a skydiver?" Mrs. A responds to this remark by adding another question to the list: "How does atmosphere relate to how a parachute works?" Mrs. A tells her students that before they build their parachutes, they will explore all of these questions. The information they gather will help them design more effective parachutes.

Developing and Using Models

Scientists develop and use models to help them understand how the world works. Engineers use models to help them design effective solutions to problems (NRC 2012, p. 58). The *NGSS* expects young children to use concrete models to communicate their findings, develop their understanding, and present their ideas (NGSS Lead States 2013).

The guiding question for the third lesson in Designing Parachutes is "How do the thickness of an atmosphere and the design of a parachute affect the speed of a falling parachute?" (Museum of Science, Boston 2011). Mrs. A begins by asking her third-grade students for their ideas about models: "What are they, and why are they important?"

Next, she shows them two plastic jars: one filled with water and a few drops of food coloring and one filled only with air. She explains that the jars model two different atmospheres, one thick and one thin. "How could we figure out which jar models which type of atmosphere?" she asks. "Feel inside the jars," one student suggests. "Drop something in each of the jars," says another. "That's what I'm going to do," Mrs. A confirms.

She holds up two golf balls, and then drops them simultaneously into the two jars. "Which jar models a thicker atmosphere?" she asks. "The one where the ball fell more slowly!" students answer. "Because there's more stuff to bump into," one girl posits. "Good thinking! The water is thicker, or denser, than the air, so it creates more drag on the ball. Which planets might this jar model?" Mrs. A asks. Students recalled their solar system knowledge: "Venus!" "Neptune!" "Jupiter!"

Mrs. A has students summarize what they learned from the models that would help them understand why parachutes fall more slowly through denser atmospheres. One boy answers, "It's like the food coloring in the water; when things drop through the atmosphere, they push through, and they move it out of the way." Models of atmospheres help students understand how the thickness, or density, of an atmosphere affects falling objects such as parachutes.

Planning and Carrying Out Investigations

Scientists plan and carry out systematic investigations to answer questions about natural phenomena. They identify what data should be recorded and what the variables are. Engineers, however, use investigations to gather data that help them specify design criteria and test their designs (NRC 2012, p. 60). The *NGSS* says that teachers should support children as they engage in these practices (NGSS Lead States 2013).

In Mrs. A's class, the children investigate how a parachute's design affects its drop speed. She has each group of students test one of three parachute features—canopy material, canopy size, and suspension line length—while keeping the other two constant. "You're going to control all of the variables except the one you're testing. What do you think that means?" she asks. "We all have to use the same thing on the bottom of our parachutes," offers a girl. "That's right," says Mrs. A. She explains that each group will drop

FIGURE 22.1

Parachute Design Data Sheet

 Canopy Size: Testing Parachutes

Directions: Construct your parachutes. Drop them three times, recording which landed first, second, and third. Then answer the questions below.

Variable	Constants	Trial 1	Trial 2	Trial 3	Observations
Small Canopy (8 inches)					
Medium Canopy (14 inches)	Material: Paper Suspension Line Length: 21 inches				
Large Canopy (18 inches)					

1. Which canopy size fell the **fastest**?_____

2. Which canopy size fell the **slowest**?_____

3. Why do you think that canopy size fell the slowest?_____

_____

its parachute from the same height, so they can compare the rates at which each falls. "Do you think we should drop each set of parachutes just once?" Mrs. A asks. "No!" shout several children. "Why not?" she probes. "It might not drop right the first time; you want to make sure it's right," says a boy.

Each group of students creates three parachutes for the investigation, changing one variable and controlling the others. For example, the "canopy materials" groups build three parachutes using three different canopy materials, while keeping canopy size and suspension line length constant.

Mrs. A then shows the class how to record their data. The students simultaneously release their three parachutes from the same height and record the order in which they hit the ground (see Figure 22.1). Each group conducts three trials of the experiment and writes the results for each trial on a recording sheet. Once all groups have completed their tests, Mrs. A has them share their findings with the class. She records everyone's findings on a data table on chart paper, so everyone can see. Therefore, even though each group focused on a single parachute variable, all students now have access to a larger data set that they can use to draw conclusions about parachute features and drop speed.

Analyzing and Interpreting Data

Scientists analyze and interpret data to generate evidence for scientific theories. Engineers analyze and interpret data to better understand design flaws and strengths and how they can be improved. Our elementary students engage in these practices as they collect data, tabulate or graph it, and share it with the class. The teacher supports them in these processes, especially in sharing and interpreting data, as the *NGSS* specify (NGSS Lead States 2013).

After conducting the investigations and sharing their data, Mrs. A's students discuss their findings. "Why did the large canopy work better?" Mrs. A asks. "It had to push more stuff out of the way as it fell," one girl answers. "It's like the jar with water." "The small parachute doesn't open that big, so it falls fast," says another. They've seen clear patterns in the data and are able to posit explanations about what's happening.

The students need more help interpreting the investigations that compared different lengths of suspension lines. Mrs. A asks, "What happened when we dropped a parachute with really short suspension lines?" "It didn't open that much; it fell like a rock!" a boy answers. "Longer lines let the parachute open, so it goes slower," argues another boy. "There's more air resistance and drag," Mrs. A clarifies. Students will use what they learn from this class discussion within their

FIGURE 22.2

Student Parachute Plans

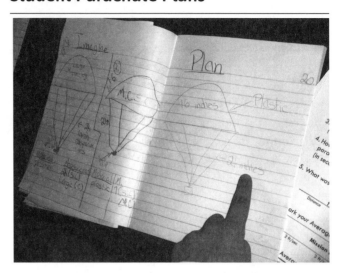

teams as they juggle the benefits and trade-offs of each design feature to create parachutes that not only fall slowly, but also fit on a spacecraft.

Engaging in Argument From Evidence

Scientists engage in argument from evidence to test and strengthen their ideas. Engineers engage in argument to compare and strengthen their designs (NRC 2012, p. 73). According to the *NGSS*, elementary students should practice advocating for their own ideas. They need the teacher's support to learn the difference between opinion and evidence-based argument (NGSS Lead States 2013).

To help students engage in argumentation before they create and test their parachute designs, Mrs. A challenges students to individually brainstorm at least two different parachute designs, taking into consideration canopy size, canopy material, and suspension line length (see Figure 22.2). Students then share their ideas with their groups.

Mrs. A encourages groups to combine ideas from multiple students into their designs. She also reminds them to use their data on how design variables affect drop speed to inform their decision making.

"So what's your team thinking?" Mrs. A asks one group. "He thinks we should use coffee filters for our canopy," a girl replies. "But we saw they have holes in them, so air can go through," says one of her group members. "What do the rest of you think?" Mrs. A asks. "Remember to look back at your data." The students check their notebooks. "The plastic bag fell the slowest; I think we should use that," one boy says.

By reminding the group to rely on the information collected during prior investigations, Mrs. A helps develop students' capabilities to consider teammates' ideas, argue from data and evidence, and compromise when selecting design ideas—skills that are central to engineering design (performance expectation 3-5-ETS1-2: "Generate and compare multiple possible solutions to a problem based on how well each is likely to meet the criteria and constraints of the problem"; NGSS Lead States 2013).

Constructing Explanations and Designing Solutions

To meet the goals of both science and engineering, the *NGSS* requires that elementary students have opportunities to construct, test, and evaluate their understanding of the world (based on their observations) and to design, test, and evaluate solutions to problems (disciplinary core idea ETS1.B: Developing Possible Solutions; NGSS Lead States 2013).

We've already presented one example of how students construct explanations in the section "Analyzing and Interpreting Data." Here is another. As each group drops its parachute in the stairwell of the school, Mrs. A times the descent (while also making sure that students behave safely while in the stairwell). One team's

parachute crashes to the floor. Their short suspension lines prevented the canopy from opening. "What do you think the problem is?" Mrs. A asks. "Could be the suspension lines," one child answers, "or the canopy might be too small." As they test and observe their parachutes and share their findings with the class (see Figure 22.3), groups construct an understanding of how parachutes work and how different design elements impact parachute performance.

FIGURE 22.3

Parachute Testing Results

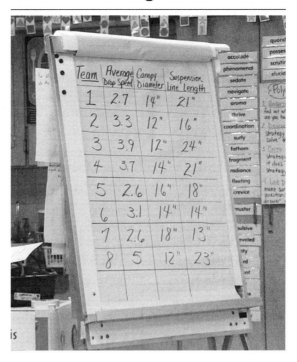

Obtaining, Evaluating, and Communicating Information

Scientists and engineers obtain, evaluate, and communicate information through scientific texts, graphs, and data and by evaluating presentations and design prototypes (NRC 2012, p. 76). The *NGSS* advocate that the process of learning to obtain, evaluate, and communicate information begin in the earliest grades, as children read age-appropriate science texts, write in science journals, and create design plans for themselves, their peers, and their teachers (disciplinary core idea ETS 1.B: Developing Possible Solutions; NGSS Lead States 2013).

Using Mathematics and Computational Thinking

Mathematics and computation are vital to both engineering and science. They enable the communication of precise ideas and the ability to make inferences and draw conclusions from data (NRC 2012, p. 66). The *NGSS* say that elementary students should begin engaging in these practices by making measurements, identifying patterns in data sets, and mathematically describing data sets using simple statistics (NGSS Lead States 2013).

In the Designing Parachutes unit, students measure and cut circles for parachute canopies and string for suspension lines, find the mean of three timed trials, and calculate the mean rate at which their parachutes fall (see Figure 22.4). They round decimals and learn when and why rounding is appropriate, putting their math skills into practice.

FIGURE 22.4

Parachute Design Data Sheet

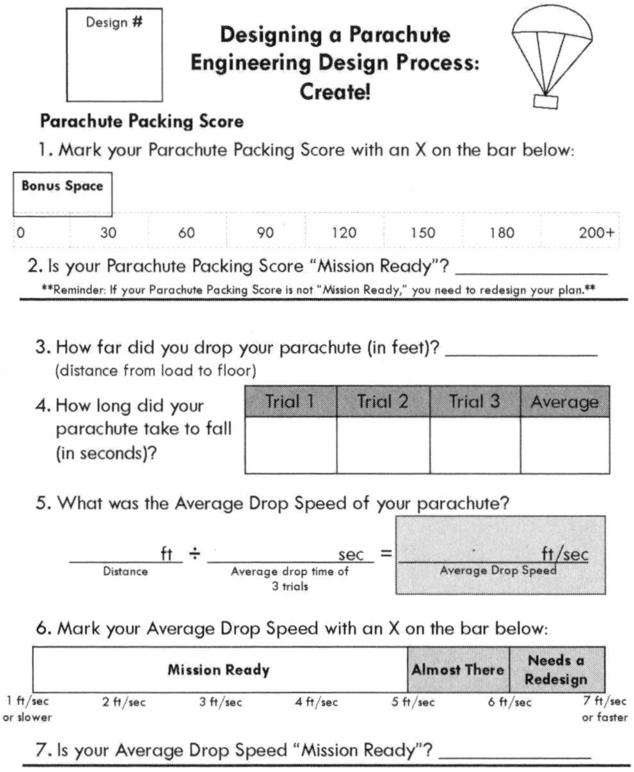

Design #

Designing a Parachute
Engineering Design Process:
Create!

Parachute Packing Score

1. Mark your Parachute Packing Score with an X on the bar below:

Bonus Space

0 | 30 | 60 | 90 | 120 | 150 | 180 | 200+

2. Is your Parachute Packing Score "Mission Ready"? ______________

Reminder: If your Parachute Packing Score is not "Mission Ready," you need to redesign your plan.

3. How far did you drop your parachute (in feet)? ______________
(distance from load to floor)

4. How long did your parachute take to fall (in seconds)?	Trial 1	Trial 2	Trial 3	Average

5. What was the Average Drop Speed of your parachute?

______ ft ÷ ______ sec = ______ . ______ ft/sec
Distance — Average drop time of 3 trials — Average Drop Speed

6. Mark your Average Drop Speed with an X on the bar below:

Mission Ready	Almost There	Needs a Redesign

1 ft/sec or slower | 2 ft/sec | 3 ft/sec | 4 ft/sec | 5 ft/sec | 6 ft/sec | 7 ft/sec or faster

7. Is your Average Drop Speed "Mission Ready"? ______________

Value of Engineering Beyond Science

Engineering asks students to apply their science knowledge in meaningful ways and to engage in engineering practices. However, the value of engineering in classrooms goes far beyond this intersection of subject areas. By tackling engineering design challenges, students practice 21st-century skills (see Internet Resource, p. 187) such as creativity, collaboration, critical thinking, and problem-solving. Engineering design challenges are also inherently open-ended, with many possible design solutions. The open-ended nature of engineering design challenges encourages students to tap into their creativity and think outside of the box. Working with their peers, students communicate their design ideas and collaborate with teammates to arrive at a final design. Finally, at its core, engineering is a problem-solving process. As students gain experience with engineering and science practices, they also become more effective—and creative—problem solvers.

Connecting to the *Next Generation Science Standards*

The materials, lessons, and activities outlined in this chapter are just one step toward reaching the performance expectations listed below. Additional supporting materials, lessons, and activities will be required.

3-5-ETS1 Engineering Design *www.nextgenscience.org/3-5ets1-engineering-design* **2-PS1 Matter and Interactions** *www.nextgenscience.org/2ps1-matter-interactions*	**Connections to Classroom Activity**
Performance Expectations	
3-5-ETS1-2: Generate and compare multiple possible solutions to a problem based on how well each is likely to meet the criteria and constraints of the problem	Brainstormed ideas for a parachute design and compared ideas with others in the group, drawing on what they have learned about how canopy size, material, and line length affect drop speed Evaluated how likely each idea is to meet the criteria and constraints of the problem
3-5-ETS1-3: Plan and carry out fair tests in which variables are controlled and failure points are considered to identify aspects of a model or prototype that can be improved	Conducted tests and generated data comparing different canopy materials, canopy size, and suspension line length
2-PS1-2: Analyze data obtained from testing different materials to determine which materials have the properties that are best suited for an intended purpose	Collected and tabulated data concerning types of materials best suited to make a parachute and share with the class so it can be analyzed to inform future designs
Science and Engineering Practices	
Asking Questions and Defining Problems	Asked questions to define the problem of how to design a parachute that will put a payload on another planet Identified questions to be investigated about how parachutes work
Developing and Using Models	Used a model to describe what happens when a parachute falls through atmospheres of different thicknesses on other planets
Planning and Carrying Out Investigations	Conducted a collaborative fair test of variables affecting the drop rate of parachutes, producing data to be used by the class to make decisions about how to plan a design solution
Analyzing and Interpreting Data	Tabulated the data collected by different groups and displayed it for the class for analysis to determine how the variables affect parachute drop speed, thus informing design
Engaging in Argument From Evidence	Debated the relative merits of different parachute designs, drawing on evidence collected earlier

3-5-ETS1 Engineering Design www.nextgenscience.org/3-5ets1-engineering-design 2-PS1 Matter and Interactions www.nextgenscience.org/2ps1-matter-interactions	Connections to Classroom Activity
Science and Engineering Practices	
Constructing Explanations and Designing Solutions	Constructed an explanation of what the model atmospheres could illustrate about how atmospheres work and explanations of why different parachutes perform differently when testing certain criteria
Obtaining, Evaluating, and Communicating Information	Communicated parachute design plans using diagrams with labeling and specification of key features
Using Mathematics and Computational Thinking	Measured and cut circles for parachute canopies and string for suspension lines and parachute drop times, analyzed data collected concerning parachute drop, and used this data to construct improved models
Disciplinary Core Ideas	
ETS1.A: Defining and Delimiting Engineering Problems • Possible solutions to a problem are limited by available materials and resources (constraints). The success of a designed solution is determined by considering the desired features of a solution (criteria). Different proposals for solutions can be compared on the basis of how well each one meets the specified criteria for success or how well each takes the constraints into account.	Defined constraints and criteria and identified those relevant to the challenge to design a parachute for another planet Developed a variety of proposals for solutions and compared them, each group choosing one design to create and test against criteria Analyzed tests of the prototype to develop improvements and further test
ETS1.B: Developing Possible Solutions • Research on a problem should be carried out before beginning to design a solution. Testing a solution involves investigating how well it performs under a range of likely conditions. • At whatever stage, communicating with peers about proposed solutions is an important part of the design process, and shared ideas can lead to improved designs.	Conducted tests of materials in collaborative groups and engaged in an investigation of the effects of atmosphere thickness before beginning project Tested a variety of materials to determine which is best suited to make a parachute to fulfill specific needs Shared and debated ideas in groups, developed design solutions, and then tested for drop speed and packing size
ETS1.C: Optimizing the Design Solution • Different solutions need to be tested in order to determine which of them best solves the problem, given the criteria and the constraints.	Compared parachute prototypes after testing and scored them against criteria and constraints to determine which designs best solve the problem
PS1.A: Structure and Properties of Matter • Different properties are suited to different purposes.	Tested a variety of materials to determine which is best suited to make a parachute to fulfill specific needs

3-5-ETS1 Engineering Design *www.nextgenscience.org/3-5ets1-engineering-design* 2-PS1 Matter and Interactions *www.nextgenscience.org/2ps1-matter-interactions*	**Connections to Classroom Activity**
Crosscutting Concept	
Influence of Science, Engineering, and Technology on Society and the Natural World	Discussed the use of parachutes to drop scientific instruments on Mars, the possible use of parachutes on other planets, and the way engineers help develop new kinds of parachutes for these new conditions

Source: NGSS Lead States 2013.

References

Museum of Science, Boston. 2011. *A long way down: Designing parachutes.* Boston, MA: Engineering is Elementary.

National Research Council (NRC). 2012. *A framework for K–12 science education: Practices, crosscutting concepts, and core ideas.* Washington, DC: National Academies Press.

NGSS Lead States. 2013. *Next Generation Science Standards: For states, by states.* Washington, DC: National Academies Press. *www.nextgenscience.org/next-generation-science-standards.*

Internet Resource

Partnership for 21st Century Learning *www.p21.org*

Designing a Sound-Reducing Wall

Students Explore Engineering With a Fun Challenge

By Kendra Erk, John Lumkes, Jill Shambach, Larry Braile, Anne Brickler, and Anna Matthys

Many of the structures and environments that we encounter in our daily lives are designed to interact with sound waves in a very specific way. For example, the carpeted floors of a library or movie theater are designed to absorb sound waves and create a quiet environment, although the smooth tile floors of a gymnasium are designed to reflect sound waves and create a noisier environment. Acoustic engineers use their knowledge of sound to design quiet environments (e.g., classrooms and libraries) and design environments that are supposed to be loud (e.g., concert halls and football stadiums). They also design sound barriers, such as the walls along busy roadways that decrease the traffic noise heard by people in neighboring houses.

To help students better understand sound in a fun and engaging way, we conducted an engineering design–based science learning activity that is appropriate for a one-hour time block in a third- or fourth-grade classroom. In this activity, student teams were challenged to design, construct, and test a sound-reducing wall created from common classroom materials. Before we

began the activity, we described the basic properties of sound waves to the students, including the ability of sound waves to travel through the air from the sound source (e.g., alarm clock) to a nearby listener (e.g., sleeping person) and their ability to be reflected or absorbed by different materials (see sidebar, "The Science of Sound"). This activity allowed students to explore the sound-absorbing or sound-reflecting properties of different types of materials, providing connections between science concepts and real-world experiences. It could be useful to perform this engineering design–based activity following an inquiry-based activity that explores the properties of sound in-depth, including the transmission of sound by waves and vibrations (Merwade et al. 2014). In the sound-reducing wall design activity, the participating students demonstrated the ability to articulate the key aspects of solving a problem using the design process and at the same time were excited about implementing, assessing, and communicating the results of their team effort. It is important to note that results are described from a grade 3 classroom because historically the science of sound was a third-grade academic science standard in the state of Indiana. Additionally, there are opportunities to extend this activity through connections to the *Common Core State Standards, Mathematics*, particularly with recording, graphing, and interpreting numerical data.

The Science of Sound

When sound waves come in contact with different objects, portions of the waves are typically *reflected* and *absorbed*. For the reflected portion of a sound wave, the forward motion of the wave is changed by the object and the wave bounces back toward the source of the sound (e.g., an echo response) or in a different direction. Most objects also absorb energy from the sound wave, causing the sound level of the reflected wave to be reduced. The sound will appear to be quieter or possibly disappear entirely to a person listening nearby.

When energy from a sound wave is absorbed by an object, the absorbed energy causes the atoms in the object to rapidly vibrate and bump into neighboring atoms, allowing for the sound energy to propagate or travel through the object. Depending on what the object is made of, the energy will either travel through the object very quickly and efficiently with little absorption or it will travel slowly and inefficiently with most of the energy being absorbed by the object. For example, the sound generated by a person knocking on a wooden door will travel through the wood and can easily be heard by a person on the other side of the door. Very little sound energy is absorbed by the dense wood that is used to make the door. However, the sound generated by a person walking on a carpeted floor cannot easily travel through the porous material of the carpet. Instead, the sound energy is absorbed by the material so that, for example, a person walking on a carpeted second floor of a house cannot easily be heard by a person on the first floor.

FIGURE 23.1

Two-Room Testing Model Composed of the "Quiet" Room (left) and the "Loud" Room (right), Separated by an Empty Wall Pocket

Lesson Overview

Because of space, time, and material constraints, the sound-reducing wall created by students was scaled to the size of a standard shoe box (see Figure 23.1). The sound-reducing wall divided the shoe box into two "rooms": a "loud" room in which a sound was generated with a kitchen timer or electric buzzer and a "quiet" room, which contained a sound meter for measuring the amount of sound that passed through the wall. For the sound meter, an external microphone (an earbud cable with a microphone, packaged with most smartphones) was placed in the room and connected to a smartphone that had a sound meter app installed (e.g., dB Volume). For a well-designed sound-reducing wall, there was a significant reduction in the level of sound that was measured with the sound meter when the buzzer was on and the wall was in place, compared to the level of sound when the wall was removed.

Student teams each constructed a wall that fit into this two-room testing model using a folded section of a manila folder as a "pocket" (see Figure 23.1). The teacher created one testing box for this activity and provided each student team with an empty wall pocket and access to different fill materials. Before the activity, the teacher gathered a variety of building materials that exhibited different sound-absorbing or sound-reflecting qualities. Suggested building materials for wall construction include the following:

- Construction paper
- Foam board
- Cotton balls
- Fabric
- Bubble wrap
- Plastic sheet

The building materials were cut by the teacher to fit in the pocket. An additional benefit of using the pocket and pre-cut materials to create the wall was that the students did not need to use scissors, tape, or glue to form their building materials into a solid wall; instead, the materials that were selected for the wall based on their team design were just placed or layered within the pocket. The building materials, pockets, and two-room testing model were also able to be reused in the future.

Engineering Design Challenge

Table 23.1 summarizes the stages and timing of the engineering design challenge. This activity

TABLE 23.1

Time Allotted for the Activity

Stage	Time Allotted
Read and discuss the design brief.	10 min.
Brainstorm and individually design.	5–10 min.
Design in teams and discuss.	5–10 min.
Gather materials and construct the design.	10 min.
Test the design (6 teams).	10 min.
Conduct class discussion of results and individual reflection.	10–15 min.

relied on the Science Learning Through Engineering Design model for engineering design (Capobianco, Nyquist, and Tyrie 2013). In this model, students were grouped in teams of three or four and worked together to first identify the overall context of the design problem or challenge. This was accomplished by reading and discussing a design brief, which was a short paragraph written by a team of university faculty and practicing teachers to provide the context of the problem and the criteria that must be addressed in the engineering design. The following design brief was used for this activity:

A group of students are starting a rock band. One of the students' parents will allow the band to practice in their house but only if a sound-reducing wall is installed in the student's bedroom. The parents hire the Silence Is Golden Company to design the sound-reducing wall. As one of the company's acoustic engineers, you and your team must design, build, and test a wall that reduces as much of the noise that escapes from the student's bedroom as possible. The wall should be no thicker than 4.5 cm.

TABLE 23.2

Design Brief Discussion Questions

Problem	Need a wall to reduce sound of instruments
Goal	To design a sound-reducing wall
Client	Parents
User	Parents and students in the band
Criteria	Wall should reduce sound as much as possible.
Constraints	1. Thickness of no more than 4.5 cm 2. Time 3. Building materials

After reading the design brief together as a class, the teacher helped students identify the problem, goal, client, user, criteria, and constraints, which were identified and written on the board and copied into the student's notebooks (see Table 23.2). The teacher then showed the two-room testing model, the wall pocket, and the available building materials to the students and described how the walls would be tested using the buzzer and sound meter. At this point, if the students are unfamiliar with the concept of "design" and the role of the design brief, sketching, building, and testing within the design process, additional time should be allotted to discuss the different aspects of the design brief listed in Table 23.2 (focusing on the goal,

FIGURE 23.2

Individual Design Sketch (left) and Team Design Sketch (right)

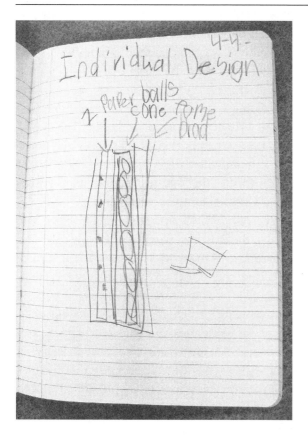
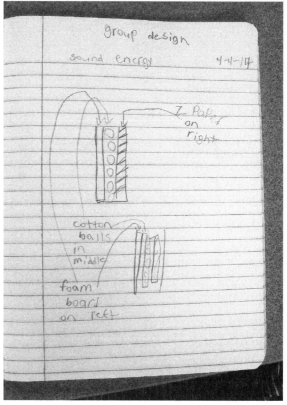

criteria, and constraints) and the brainstorming and sketching time that will follow.

Next, time was allotted for the students to individually brainstorm and sketch design ideas in their lab notebooks based on the information provided and their relevant background knowledge. The students then shared their design ideas with the members of their team and mutually agreed on a final "team design," which was then copied into the lab notebooks of all the team members. Before the teams were allowed to begin construction of their design, the teacher checked to see that each team member's lab notebook contained a sketch of the team design. Examples of students' sketches are shown in Figure 23.2. The students were told that a good sketch is neatly drawn, is large and centered on the page, and includes arrows and clearly written labels.

When the teams were ready to begin building, one student from each team gathered the materials to be used in construction of the wall based on their final team design. The teams then built the wall together using their team design as a guide and ensuring that the thickness of the wall did not exceed 4.5 cm. Constructing the teams' walls took a relatively short amount of time (only 10 minutes) because the teacher assembled the pre-cut materials.

Before testing any student-designed walls, a sound level baseline was established. To do this, an empty wall pocket was placed into the two-room testing model, the buzzer was turned on, and the lid of the box was tightly shut. The value displayed on the sound meter was the baseline sound level, and this value was recorded on the chalkboard. Next, one-by-one, the teams tested their finished walls in the two-room testing model, with help from the teacher in a similar fashion, by replacing the empty wall pocket with their team's wall. To make the testing process more interactive, a document camera was used to project the image of the smartphone sound

meter on the screen so that the students could participate in reading the measurement of the sound level. The value of the sound level for each wall was recorded on the chalkboard; sample data is shown in Table 23.3. The students copied these results into their lab notebooks. Students can also include a third column where they calculate the difference between the baseline sound level and the sound level with a team's wall; for example, for Team 1 the difference would be 11 decibels (86 − 75 = 11).

TABLE 23.3

Results From Testing the Teams' Sound-Reducing Walls

Test	Sound Level (decibels)
Baseline	86
Team 1	75
Team 2	82
Team 3	78

Following testing, the teacher led a group discussion about the results. First, the teacher asked the teams with the highest and lowest sound levels in the quiet room to describe the materials that they used to construct their wall. Those materials were listed on the board, and the students were asked why the materials make good or poor sound reducers. To encourage the students to consider how sound travels through the different materials, ask the students which materials were the most porous (or least dense) and how well the low-density materials worked to reduce the sound level in the quiet room. Also, ask whether any of the materials seemed to be particularly good at reflecting the sound energy. In their notebooks, the students ranked the materials from the most to least effective at reducing the sound level by absorption or reflection of the sound. If time permits, the teams can redesign their walls, using their new knowledge of good and poor sound-reducing materials to create a more effective sound-reducing wall. At the end of the activity, time was allotted for "self-reflection," in which the students copied and provided answers to the following five questions in their notebooks: (1) What did I do? (2) What worked well? (3) What didn't work well? (4) What would I change? and (5) What did I learn?

When this activity was implemented in the third-grade class, the students understood and were able to articulate the need, goal, client, user, criteria, and constraints after hearing the design brief. The students became very engaged in the design and testing process. When measuring each team's wall, the entire class was captivated and glued to the video screen to see the sound meter and cheered for each other when a team's wall did well. Making a list of values for each team provided friendly competition. Even teams whose walls did not reduce the sound as much as other teams were happy to share their design, talk about why they chose the materials they did, and discuss how they thought they could improve their design.

When discussing as a group why they thought each material would or would not work well in reducing the sound levels in the quiet room, students frequently exclaimed that they believed "thicker walls would work better" at reducing the sound levels and that "since cotton balls are the thickest [material], they will work the best." These predictions are understandable based on the students' past experiences—for example, putting a pillow over your head during a thunderstorm works better at blocking out the loud thunder than a thin bed sheet. Most teams attempted to construct the thickest wall possible, which even caused some of the students to abandon their team designs. For example, instead of layers of foam

board, the students chose to construct the wall from layers of cotton balls, which were thicker.

But the students quickly discovered that cotton balls did not do the best job of reducing the sound. Instead, they directly observed that thinner walls made of foam board and plastic sheet were the most effective at reducing the sound level in the quiet room compared to thicker walls made from fabric or cotton balls. This is because, for effective sound reduction, the sound waves can either be reflected or absorbed. Many soft, porous materials (such as cotton balls and fabric) absorb sound but do not reflect, whereas hard, flat materials (such as plastic sheet) reflect well but do not absorb much sound. Foam board is an example of a material that both reflects sound (off its flat, stiff surface) as well as absorbs sound (within its porous foam core). In this situation, the teacher could lead a discussion on how thicker walls may not always be better for sound reduction; the type of material that the wall is composed of matters more. If time allows, the teams could create a new team design for a sound-reducing wall based on the outcomes from the first round of testing and the posttesting discussion. This extension will allow the students to directly apply what they learned from the design activity and will also assess their overall level of understanding of how sound travels and interacts with different materials.

Assessment

The teacher performed informal assessments throughout the design process while circulating the room to observe the students' design notebooks, listened to team discussions, and questioned design plans in relation to science concepts. The teacher reviewed student notebooks for design brief details, individual design sketches, and team design sketches. The teacher provided a separate rubric for students to self-check their engineering design notebook (see NSTA

Connection, p. 197). The teacher then used the rubric to assess the engineering design notebook (see NSTA Connection). The teacher assessed the students' notebooks for the design brief information, individual designs, team designs, and student self-reflections. The teacher also administered a formal assessment after the design task to assess students' understanding of the science concept as well as the engineering design process. This assessment could be performed as a pretest and posttest (see NSTA Connection, p. 197). Additionally, the students' understanding could be assessed by having the students create a sketch or fully sketch and construct a redesigned wall based on their observations from the outcomes of the activity. Students whose redesigned walls are more effective at reducing the sound level in the quiet room than their initial walls would be categorized as having a good understanding of the properties of sound. For a simple sketch (if there is not time for a full redesign activity), students who included in their sketched walls a mix of materials that have both good sound absorption (cotton balls) and sound reflection (plastic sheet, foam board) abilities would be categorized as having a good understanding of the properties of sound.

Overall, this science activity exposes elementary students to the engineering design process of brainstorming, planning, building, and testing while simultaneously demonstrating how sound waves travel and interact with different types of materials. This fun, hands-on activity allows the students to become acoustic engineers for the afternoon, hired by a client to design and build a sound-reducing wall with certain criteria and constraints. Having a wide selection of building materials to choose from for their wall designs encouraged student creativity. At the same time, the actual construction process is simplified by using pre-cut materials provided by the teacher and a uniform testing model, allowing for all students to fully participate in the building and testing process. Working in small groups also increases the students' confidence and comfort with sharing ideas and combining separate ideas to lead to improved designs.

Connecting to the *Next Generation Science Standards*

The materials, lessons, and activities outlined in this chapter are just one step toward reaching the performance expectations listed below. Additional supporting materials, lessons, and activities will be required.

3-5-ETS1 Engineering Design www.nextgenscience.org/3-5ets1-engineering-design	Connections to Classroom Activity
Performance Expectations	
3-5-ETS1-1: Define a simple design problem reflecting a need or a want that includes specified criteria for success and constraints on materials, time, and cost	Analyzed the design brief to determine the problem, goal, client, user, criteria, materials, and constraints
3-5-ETS1-3: Plan and carry out fair tests in which variables are controlled and failure points are considered to identify aspects of a model or prototype that can be improved	Constructed, measured, and tested sound-reducing walls Recorded and analyzed sound-level data Redesigned their walls

3-5-ETS1 Engineering Design www.nextgenscience.org/3-5ets1-engineering-design	Connections to Classroom Activity
Science and Engineering Practices	
Asking Questions and Defining Problems	Analyzed the design brief
Planning and Carrying Out Investigations	Created individual sketches of their ideas Collected and analyzed quantitative results as a group
Developing and Using Models	Chose a model and constructed and tested it
Constructing Explanations and Designing Solutions	Determined which materials were the best based on their criteria
Disciplinary Core Ideas	
ETS1.A: Defining and Delimiting Engineering Problems • Possible solutions to a problem are limited by available materials and resources (constraints). The success of a designed solution is determined by considering the desired features of a solution (criteria). Different proposals for solutions can be compared on the basis of how well each one meets the specified criteria for success or how well each takes the constraints into account.	After testing and class data analysis, given the option to redesign their wall Following testing, discussed all the results from the teams' walls Identified materials that were the most and least effective in reducing sound
ETS1.B: Developing Possible Solutions • At whatever stage, communicating with peers about proposed solutions is an important part of the design process, and shared ideas can lead to improved designs.	Discussed individual design sketches in teams Created a final team sketch Constructed and tested the model Described the design to the class, including discussing why they think it worked (or didn't work)
PS3.A: Definitions of Energy • Energy can be moved from place to place by moving objects or through sound, light, or electric currents	Directly observed the results of the transfer of sound from one space to another through different materials used to reduce sound
Crosscutting Concept	
Influence of Science, Engineering, and Technology on Society and the Natural World	Improved existing technology to reduce the sound level in the quiet room

Source: NGSS Lead States 2013.

Acknowledgment

This project was supported by the National Science Foundation, Award #0962840. Any opinions, findings, and conclusions or recommendations expressed in this material are those of the authors and do not necessarily reflect the views of the National Science Foundation.

References

Capobianco, B. M., C. Nyquist, and N. Tyrie. 2013. Shedding light on engineering design. *Science and Children* 50 (5): 58–64.

Merwade, V., D. Eichinger, B. Harriger, E. Doherty, and R. Habben. 2014. The sound of science. *Science and Children* 51 (2): 30–36.

NGSS Lead States. 2013. *Next Generation Science Standards: For states, by states.* Washington, DC: National Academies Press. *www.nextgenscience.org/next-generation-science-standards.*

NSTA Connection

Visit *www.nsta.org/SC1509* for the assessment and answer key, the rubric, and additional resources.

Blade Structure and Wind Turbine Function

Third and Fifth Graders Co-investigate and Co-design Wind Turbine Blades and Voltage Output

By Pamela Lottero-Perdue, M. Angela De Luigi, and Tracy Goetzinger

Investigating and designing technologies, objects, or "built systems" allows students to engage in engineering practices and engineering design. It also enables them to gain direct evidence to support the relationship between structure and function, one of the seven crosscutting concepts within *Next Generation Science Standards* (NGSS Lead States 2013).

In this chapter, we feature the study and design of wind turbine blades by third and fifth graders at Darlington Elementary School, a small, rural school with one classroom of students per grade level. Working in same-grade groups, the students conducted controlled experiments to investigate how the structure and number of wind turbine blades affected the turbine's function. After sharing results across groups and grades, mixed-grade teams then used knowledge gained from the investigations to engineer effective and efficient wind turbine blade systems (see the "Wind Turbine Power, Energy, and Voltage" sidebar, pp. 202–203).

Context

The wind turbine lessons featured in this chapter emerged from our desire to help students build on their learning about energy transformation, which is already a part of the existing curriculum. Throughout the year, Darlington Elementary had embarked on a whole-school effort to embed energy concepts in the curriculum and special events, including a trip to the local science center. This effort extended existing third- and fifth-grade science-engineering integrated curricular units that specifically addressed energy transformation. Earlier third-grade lessons focused on the relationship between force and motion and on energy types and transformation. Then, students designed a simple windmill blade system to spin the windmill's axle, which, in turn, lifted a cup full of weights (Engineering is Elementary 2009b; Lottero-Perdue, Lovelidge, and Bowling 2010). Earlier fifth-grade lessons involved the design of an electrical circuit that signaled when a cup was emptied (Engineering is Elementary 2009a). This fifth-grade engineering design experience was set in the context of science learning about electrical energy and energy transformation, electricity, and electromagnetism.

We, the authors, came together as enthusiastic co-conspirators, eager to connect what students had already learned in these energy units to the schoolwide energy theme via a new, cross-grade, science-engineering integrated experience that would encourage collaborative teamwork. What emerged was an experimental investigation followed by an engineering design challenge to design efficient wind turbine blade systems. This investigation was inspired by the KidWind Project curriculum (see Internet Resources, p. 210) and used the KidWind turbine base, generator, and hub on which to attach blades. The entire unit lasted five instructional hours spread across five days; two-and-a-half hours were dedicated to the experimental investigation and two-and-a-half hours were dedicated to the engineering design challenge.

Introduction to the Wind Turbine

To introduce the students to the concept of a wind turbine on the first day of instruction for the wind turbine investigation, we asked students to recall their aforementioned third-grade windmill design challenge. The third graders had undertaken this challenge quite recently, and the fifth graders had taken the challenge two years earlier when they were in third grade. We asked questions such as, "What caused the windmill blades to spin?" (the wind); "What did that spinning allow the windmill to do?" (lift the cup); and "What was the intended output energy of the windmill?" (mechanical). We then showed the students the model wind turbine with simple blades affixed to it. "Instead of lifting a cup or pumping a well," we said, "the intent of a wind turbine is to transform mechanical energy into another form of energy." Holding the two ends of the wires that stretched from the body of the

turbine, we asked a bit leadingly, "What might this form of energy be?" (electrical). This discussion reinforced disciplinary core ideas about the conservation of energy and energy transfer (NGSS Lead States 2013). We then demonstrated how we would connect the wires to a digital multimeter, which would read voltage—a term familiar to both third and fifth graders in their everyday experiences with batteries. Although voltage is not the same as energy, we shared that we would use voltage to indicate the amount of electrical energy produced: the higher the voltage, the higher the electrical energy output (see the "Using Digital Multimeters to Measure Output Voltage" sidebar, p. 204).

Experimental Investigation

After the introduction to the wind turbines, day 1 continued with the first step of the investigation, which was to begin to plan the investigation by considering what independent blade-related variables might affect the turbine's voltage output—the dependent variable—when the turbine was placed directly in wind produced by a fan. Both the third and fifth graders, in separate discussions, suggested the following independent variables: angle of blades, spacing of blades, number of blades, size of blades, shape of blades, weight of blades, and blade material.

The second step of the investigation involved forming student groups and assigning, or allowing groups to choose, their independent variables. In both the third- and fifth-grade classrooms, students worked in teams of three or four; these were same-grade teams that were otherwise heterogeneous with respect to academic performance. Although the classes had no English language learners, both classes did have special needs students. Those students were placed in groups where they would feel most comfortable to share their ideas and where peers could provide support with regard to documentation activities.

We guided the third-grade groups to investigate the simplest of the independent variables—length, width, or number of blades—while keeping the cardboard material, rectangular shape, and 20° angle constant. Note that for the angle measurement, 0° represents the flat surface of the blade being perpendicular to the wind, and 90° is parallel to the wind.

We allowed the fifth-grade groups to select the independent variable they would investigate. The fifth-grade groups explored a wider range of variables than the third -grade groups, including the angle of blades, blade shape, and blade orientation. Blade shape referred to the overall shape of the blade, such as comparing a triangular or rectangular shape while keeping surface area constant. Another shape group was curious about the shape of the edge: Does a wavy, straight, or jagged edge produce higher voltage? Blade orientation referred to the way in which a certain-shaped blade was mounted to the dowel. The orientation team asked: How does the orientation of a square—either as a rhombus or a square (see Figure 24.1)—affect the voltage produced?

Day 2 of the unit began with the third step of the investigation when students worked collaboratively in their groups to plan their investigations. Students used planning sheets and white boards to sketch ideas about measurements and numbers of blades. Some fifth-grade teams created pattern pieces using scissors, paper, and rulers to further plan their designs. Groups determined how they would alter their independent variable three times in their experiment. For example, the "number of blades" group determined that they would measure voltage output for three, four, and six blades, respectively. Each group then predicted in a whole-class discussion how output voltage would be affected by changing the experimental variable.

Safety Tips

- Urge students to use caution when cutting cardboard with scissors, helping them as necessary.
- Teachers or other adults should use hot glue to affix blades firmly to dowels.
- A teacher or other adult should operate the fan. Students should wear safety goggles and stand back approximately 0.5–1 meter from the testing station.

The fourth step of the investigation was to construct the blades. Materials for each group included scissors; three to six 30 cm × 50 cm cardboard rectangles that had been precut by teachers from packaging boxes; one 18 cm long 0.6 cm (¼ in.) dowel per blade; two 5 cm pieces of masking tape per blade; and one hub, which had multiple holes into which the dowels could fit and another hole that could fit securely onto the shaft of the wind turbine. Additionally, the teacher or other adult assistant at the hot glue station had a hot glue gun and glue sticks. Groups carefully

FIGURE 24.1

Rhombus vs. Square Orientation of a Wind Turbine Blade

Black dot represents axis of turbine

Direction of spin shown in gray

Rhombus Orientation

Square Orientation

FIGURE 24.2

Affixing the Hub to the Turbine Axle

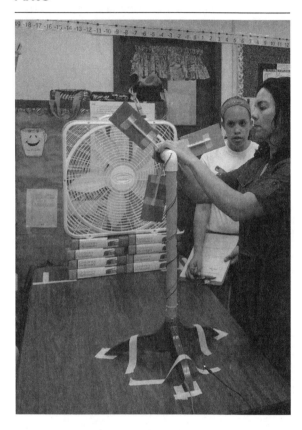

cut and assembled their blades and affixed them to dowels with tape. They brought these assembled blades to the hot glue station for teachers to permanently affix the dowels to the blade. Then, they inserted each blade dowel into the turbine hub (see the "Safety Tips" sidebar, p. 201).

The fifth step of the investigation was to test the blade system at the wind turbine station. The turbine station included the wind turbine base, digital multimeter, and fan. The KidWind Basic Wind Experiment Kit, including the base and generator, costs approximately $100. This basic kit also includes a special protractor to measure blade angle, 25 dowels to use to affix blades to the hub, balsa wood and polymer-based cardboard pieces (for blade materials), and supplies

to transform the wind turbine into a windmill that lifts a cup (see Internet Resources, p. 210). A set of five hubs costs $35. A simple digital multimeter costs approximately $10. All items can be purchased through *www.kidwind.org*. A box fan is recommended for the testing station; these fans can be purchased at home improvement stores for approximately $20.

Each group repeated the testing step three times as students re-configured their hub-and-blade systems for each version of their independent variable. At the experimental station, the teacher affixed the hub to the turbine axle and turned on the fan to the same predetermined setting; students stood back to observe the blades in motion and watch the digital multimeter (see Figure 24.2 and the "Safety Tips" sidebar).

The sixth step of the investigation involved students discussing, making sense of, and documenting their results in their groups and with their teachers. Students not only recorded their voltage data on group data sheets (see NSTA Connection, p. 210), but also recorded a claim and wrote statements to support that claim using evidence (i.e., engaging in argument from evidence). We assisted students as they tried to make sense of their data, asking questions such as, "What happened to voltage output when the length of the blades was increased?" and "Did you see a difference in the voltage output each time you increased the angle of the blade?"

Wind Turbine Power, Energy, and Voltage

Wind turbines convert the mechanical energy of the turning turbine blades into electrical energy through the use of a generator. Some of the mechanical energy is also converted into heat energy because of the moving parts inside the turbine. The output of full-scale, industrial wind turbines is typically measured in megawatts (MW) of *power*, where one MW is 1 million watts. *Power* is related to *energy* in that power is the rate of energy use. If an industrial wind turbine generates about 1.5 MW of power, enough to power about 250 homes, over the course of 12 hours, then the energy of this turbine is 1.5 MW × 12 hours = 18 MWh (KidWind 2012).

To get a sense of the electrical energy generated by the model wind turbines featured in this chapter, elementary students can measure the output voltage of their simple wind turbines. The voltage indicates how fast the blades and generator shaft spin. Voltage is proportional to the amount of energy produced: the greater the output *voltage*, the greater the output *power* and thus *energy*. The *power* generated by a wind turbine is actually equal to the product of the *current* and *voltage* generated by the turbine. The current is the measure of the flow rate of electrons in the turbine, and the voltage is the electric pressure that drives the current.

For more information about wind turbines and science related to wind turbines, including details about generators, go to *www.kidwind.org* (KidWind 2012).

In some cases, there was little or no difference in voltage output when a variable changed. This occurred for the fifth-grade shape group that compared straight, curvy, or jagged edging. Differences in their data were on the order of 0.01 volt. We asked the team, "Is that a large or a small difference between those voltage output amounts?" (small) and "Was the number in the hundredths place on the digital multimeter fixed, or did it shift from one number to another?" (shifted). Although we guided students to consider that the voltage differences

were insignificant, we were sure to regard their experiment as being very significant: It let us know for later, when we would don our proverbial "engineering hats," that cutting jagged or curved edges into the cardboard would not be worth the effort.

Experimental results suggested that higher voltage output was associated with the following:

- Shorter blades

- Fewer blades (e.g., two as opposed to four)

- Blades at a small angle (e.g., 10°, not 30°)

- Blades with the bulk of their area close to the axis of the turbine, which was learned from the rhombus (0.50 volts) versus square (0.68 volts) orientation experiment

One variable with an unclear relationship to output voltage was blade width. When comparing a width of 5 cm (2 in.) to a width of 13 cm (5 in.), voltage increased with width; however, voltage output amounts were the same for a width of 5 cm (2 in.) versus 8 cm (3 in.).

We discussed the somewhat complex point described in the final bullet above with the fifth graders, touching on the idea of rotational inertia. We demonstrated this concept using an adult and a rotating office chair, spinning quickly with arms held close the body and then slowing down as arms became outstretched. Just as the amount of material in the square did not change when it was re-oriented as a rhombus, the mass of the person in the chair did not change. What changed was how she distributed herself, making it more difficult to spin when more of her was held outward. Note that although we were able to introduce the concept of rotational inertia by reflecting on experimental results and the chair demonstration, the term was not emphasized.

24

Using Digital Multimeters to Measure Output Voltage

Elementary students can use a simple handheld device called a digital multimeter to measure output voltage. The two wires of the multimeter are clipped to the two wires that emerge from the model wind turbine's generator. When the turbine blades move and the multimeter is turned on and set to read voltage, voltage numbers—typically on the order of 0.1 volts to 2.0 volts—display on the multimeter. The higher the number, the greater the output voltage and output energy.

Sharing our results across grades was the final step in the investigation and was the starting point for day 3 of the unit. Each group's results were displayed in a poster format (see Figure 24.3 for the fifth-grade example); the third- and fifth-grade teams met in the gymnasium so that each group could share its posters, claims, and evidence-based reasoning to support those claims. All students documented the third- and fifth-grade team findings on a results sheet (see NSTA Connection, p. 210, and Figure 24.4), recording notes about interesting findings and "big idea" claims. During group presentations, students were respectful listeners and curious about what others, whether they were younger or older, had learned about how the structure of wind turbine blades affected electrical energy output.

FIGURE 24.3

A Fifth-Grade Group's Results

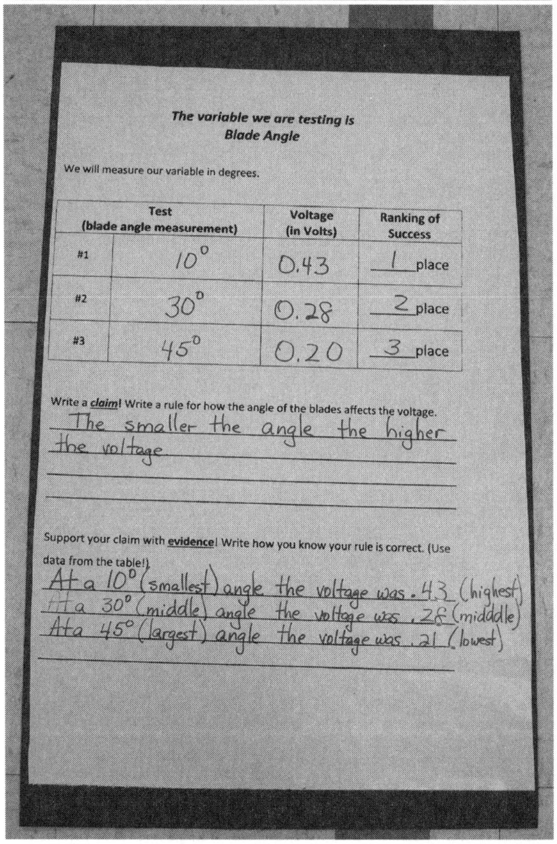

The variable we are testing is
Blade Angle

We will measure our variable in degrees.

Test (blade angle measurement)		Voltage (in Volts)	Ranking of Success
#1	10°	0.43	1 place
#2	30°	0.28	2 place
#3	45°	0.20	3 place

Write a ***claim***! Write a rule for how the angle of the blades affects the voltage.
The smaller the angle the higher the voltage

Support your claim with **evidence**! Write how you know your rule is correct. (Use data from the table!)
At a 10° (smallest) angle the voltage was .43 (highest)
At a 30° (middle) angle the voltage was .28 (middle)
At a 45° (largest) angle the voltage was .21 (lowest)

FIGURE 24.4

Results Sheet

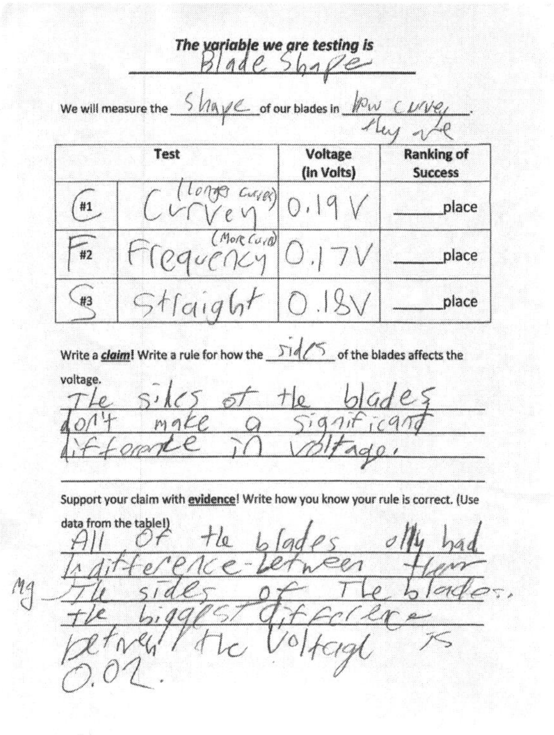

The variable we are testing is
Blade Shape

We will measure the shape of our blades in how curve they are.

	Test	Voltage (in Volts)	Ranking of Success
C #1	Curved (longer curves)	0.19 V	____place
F #2	Frequency (More curve)	0.17V	____place
S #3	Straight	0.18V	____place

Write a ***claim***! Write a rule for how the sides of the blades affects the voltage.
The sides of the blades don't make a significant difference in voltage.

Support your claim with **evidence**! Write how you know your rule is correct. (Use data from the table!)
All of the blades only had a difference between them
Mg The sides of the blades.
The biggest difference between the voltage is 0.02.

FIGURE 24.5

Design Brief for the Wind Turbine Design Challenge

Problem Statement: We need to produce as much electrical energy as possible!

Goals

- To design a wind turbine blade system that produces the most voltage possible

- To apply what we learned in our science investigations to our blade design

Design Constraints (limits)

- There will be three or four people per team—a mixture of third and fifth graders.

- The blade material must be cardboard.

- Blades must be built onto pre-cut dowels.

- Blades must be able to fit on the wind turbine without scraping the turbine base or table.

- The fan setting will be placed on a level 2 wind speed.

- The placement of the turbine and fan will be the same for all tests.

Design Criteria (how we know when we're successful)

- Blades must not fall apart.

- The higher the voltage output of the turbine, the better!

Engineering Design Process

After sharing our experimental results on day 3, students were excited to use the engineering design process (EDP) to develop, compare, and test possible blade system designs. This engagement in the EDP addresses performance expectations and disciplinary core ideas with respect to engineering design (see "Connecting to the *Next Generation Science Standards*," p. 209). We presented them with the design brief, which included the goals, constraints, and criteria for the design challenge (see Figure 24.5). The goals of the challenge were to (a) design a wind turbine blade system that would produce the most amount of voltage possible and (b) apply what we had learned in our investigations to our blade designs. We reminded the students that they would be using the EDP, based on the Engineering is Elementary curriculum, which they had used in many other units of study (see Table 24.1, p. 206).

Students were placed in mixed-grade engineering design teams, most of which had two third-grade students and two fifth-grade students. As with the single-grade experimental investigation groups, mixed-grade design teams were constructed to reflect heterogeneity; special consideration was given to special needs students to ensure that their team members would be patient, helpful, and willing to listen. Teams were reminded to use what they *and their peers* had learned in the prior investigation as they designed their wind turbine blade systems. This reminder, as well as the presentation of the design brief, represented the Ask step of the

TABLE 24.1

Engineering Is Elementary Engineering Design Process Steps and Descriptions

Engineering Design Process Step	Description of Step
1. Ask	• Identify the problem. • Determine design constraints (e.g., limitations on materials that can be used). • Consider relevant prior knowledge (e.g., science concepts).
2. Imagine	• Brainstorm design ideas. • Draw and label those ideas.
3. Plan	• Pick one idea. • Draw and label the idea. • Identify needed materials or conditions.
4. Create	• Carry out the plan; create the design. • Test the design.
5. Improve	• Reflect on testing results. • Plan for, create, and test a new (improved) design.

Source: Lottero-Perdue, Lovelidge, and Bowling 2010.

EDP. For the rest of day 3, student teams worked through the Imagine and Plan steps of the EDP (see Figure 24.6). They used EDP sheets (see NSTA Connection, p. 210) to scaffold this process (science and engineering practice Constructing Explanations and Designing Solutions; NGSS Lead States 2013).

During the Imagine and Plan steps of the EDP, we constantly asked the following questions: "What does your imagined idea or plan look like?" "Why did you choose to do that?" and "What did you learn from our experimental investigations that is helping you make decisions

FIGURE 24.6

Student Teams Worked Through the Imagine and Plan Steps of the EDP

on your design?" Although many teams intentionally incorporated information from the earlier investigations to inform their designs, others (e.g., a team that used six blades in their first designs) did not. In the latter cases, if simple questioning did not nudge the teams to reconsider their designs, we allowed them to go forward; subsequent test results would present these teams with reasons to rethink their design decisions.

Day 4 of the study was the Create step of the EDP. Students engaged in this step using the same materials as had been available in the experimental investigation. Again, they were reminded of safety procedures with regard to the use of scissors and hot glue, which was applied by an adult. Once their original blade designs were constructed and latched into the hub, teams were ready to test. As students tested their wind turbine blade systems at the testing station, we asked, "What do you plan to do next to try to improve your voltage output based on these results? What did you learn from our earlier investigations (i.e., about blade structure) that might help in this improvement?" Many groups tried to improve the results by decreasing the number of blades to two while keeping the original blade design. One group chose to widen the blades based on the earlier results that indicated that wider blades were more efficient. For most teams, day 4 also included movement into the Improve step, where they not only considered how to improve their original designs, but also began to plan and create their revised designs.

On day 5 of the unit, teams implemented their Improve plans by creating and testing their revised designs, and in some cases, a third design. Figure 24.7 shows the voltage results for the first and second designs for all teams, with additional results for a third and fourth design for some teams who had time to continue the design process. Surrounding the yellow voltage results sheet are team sheets showing the shape of their initial blade design, with smaller text above each shape indicating the chosen angle and number of blades. In a final debriefing meeting, a representative from each team described to the other teams how their team had moved through the design process.

FIGURE 24.7

Voltage Results

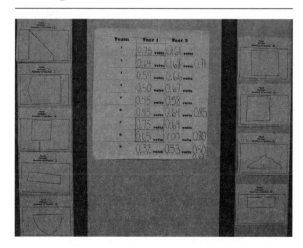

The voltage output values across all designs ranged from 0.32 volts to 1.00 volts. Of the nine teams, only three were not able to increase the voltage for the second design by 0.05 volts or more. More important than particular results, however, were the occasions when teams considered, applied, discussed, and even reconsidered their knowledge of the structure-function relationship between wind turbine blade design and resulting voltage output. One case of reconsideration had to do with blade width. Recall that experimental results regarding blade width were unclear. However, when Team 8, which started with a two-blade, wide-base triangular design, decided to try lopping off the ends to narrow each blade, the result was the highest voltage output reading for the entire challenge. The width-voltage structure-function relationship was clarified.

Assessing Student Learning

Four key questions guided our formative and summative assessment of student learning:

1. Were students' or groups' claims about a structure-function relationship supported by the investigation data they had collected?

2. Did students or teams apply structure-function investigation results (from both classes) when designing their first set of wind turbine blades?

3. Did students/teams apply investigation results (see question 2) and first design test results when designing their second set of wind turbine blades?

4. Could students articulate the way in which the wind turbines transformed mechanical energy to electrical energy (and heat energy)?

We determined the answers to these questions by constantly monitoring and reviewing student-generated documents, and through multiple, question-infused discussions with students, groups, teams and classes (see NSTA Connection, p. 210, for the rubric). We also evaluated student-generated documents for completeness and accuracy as a summative assessment of their participation and learning.

Conclusion

In one respect, the engineering design challenge made understanding the relationship between structure variables and the voltage output function of the wind turbine more complex, because multiple variables influenced each design. However, the results shared in the final debriefing meeting coalesced into the students' realization that wind turbine blades must balance two needs: (1) the ability to catch the wind to turn the axle, and (2) the ability to spin fast. Bulky designs with outstretched blades struggled to move their own weight around the axle (having too much rotational inertia); designs that used smaller blades with more surface area closer to the axis were able to move more quickly, generating more voltage output. These larger ideas, supported also by most of the experimental results, were part of the rich discussions among the third- and fifth-grade students.

We must also share that the enthusiasm of the students was palpable throughout the wind turbine investigations. Students discussed design ideas with their groups and created detailed drawings to explain their ideas—all without the need for reminders to stay "on task." They eagerly measured and cut cardboard prototypes of their ideas and argued persuasively for their designs. Although the unit occurred during the last weeks of the school year, many students and teams pleaded to be able to make "just one more" improvement. Furthermore, students worked exceptionally well together. As third- or fifth-grade groups shared experimental findings, other third- and fifth-graders listened carefully; each group had gathered information that the other groups needed. Despite the age differences on EDP teams, students were on a level playing field with respect to creativity, the ability to contribute to the design process, and yes—energy!

Connecting to the *Next Generation Science Standards*

The materials, lessons, and activities outlined in this chapter are just one step toward reaching the performance expectations listed below. Additional supporting materials, lessons, and activities will be required.

4-PS3-4 Energy *www.nextgenscience.org/4-ps3-4-energy* 3-5-ETS1 Engineering Design *www.nextgenscience.org/3-5ets1-engineering-design*	Connections to Classroom Activity
Performance Expectations	
4-PS3-4: Apply scientific ideas to design, test, and refine a device that converts energy from one form to another **3-5-ETS1-2:** Generate and compare multiple possible solutions to a problem based on how well each is likely to meet the criteria and constraints of the problem **3-5 ETS 1-3:** Plan and carry out fair tests in which variables are controlled and failure points are considered to identify aspects of a model or prototype that can be improved	Used the engineering design process to design, create, test, compare, and optimize wind turbine blade systems to maximize output voltage of a model wind turbine, which converts mechanical to electrical energy
Science and Engineering Practices	
Planning and Carrying Out Investigations	Planned and carried out investigations to determine the effect of a wind turbine blade system variable on voltage output
Engaging in Argument From Evidence	Made a claim, supported by evidence, regarding the effect of that variable on voltage output
Constructing Explanations and Designing Solutions	Engineered wind turbine blade systems to maximize voltage output
Obtaining, Evaluating, and Communicating Inform	Communicated experimental and design results within and across teams
Disciplinary Core Ideas	
PS3.B: Conservation of Energy and Energy Transfer • Energy can also be transferred from place to place by electric currents, which can then be used locally to produce motion, sound, heat, or light. The currents may have been produced to begin with by transforming the energy of motion into electrical energy.	Applied their understanding of energy transfer as they design wind turbine blade systems that turn the turbine shaft as quickly as possible (motion), which—through the turbine's generator—produces as much output voltage as possible

4-PS3-4 Energy www.nextgenscience.org/4-ps3-4-energy 3-5-ETS1 Engineering Design www.nextgenscience.org/3-5ets1-engineering-design	Connections to Classroom Activity
Disciplinary Core Ideas	
ETS1.B: Developing Possible Solutions • An often productive way to generate ideas is for people to work together to brainstorm, test, and refine possible solutions. Testing a solution involves investigating how well it performs under a range of likely conditions.	To develop possible solutions, used the engineering design process (considered the problem, constraints, and criteria) brainstormed with peers
ETS1.C: Optimizing the Design Solution • Different solutions need to be tested in order to determine which of them best solves the problem, given the criteria and the constraints.	Planned, created, tested, and optimized the wind turbine blade system design solution
Crosscutting Concept	
Structure and Function	Determined the relationship between wind turbine blade system variables and voltage output in the experimental investigation

Source: NGSS Lead States 2013.

References

Engineering is Elementary (EiE). 2009a. *An alarming idea: Designing alarm circuits.* Boston, MA: Museum of Science, National Center for Technology Literacy.

Engineering is Elementary (EiE). 2009b. *Catching the wind: Designing windmills.* Boston, MA: Museum of Science, National Center for Technology Literacy.

KidWind. 2012. Learn wind. KidWind Project. *http://learn.kidwind.org/sites/default/files/learn_wind.pdf.*

Lottero-Perdue, P. S., S. Lovelidge, and E. Bowling. 2010. Engineering for all: Strategies for helping all students succeed in the engineering design process. *Science and Children* 47 (7): 24–27.

NGSS Lead States. 2013. *Next Generation Science Standards: For states, by states.* Washington, DC: National Academies Press. *www.nextgenscience.org/next-generation-science-standards.*

Internet Resources

Engineering is Elementary: Engineering Design Process

www.eie.org/overview/engineering-design-process

KidWind

www.kidwind.org

www.vernier.com/products/kidwind/wind-energy/kits/kw-bwx

NSTA Connection

Visit www.nsta.org/sc1503 for the group data sheets, results sheet, engineering design process sheet, and rubric.

25

You and Your Students as Green Engineers

Using Creativity and Everyday Materials to Design and Improve a Solar Oven

By Tess Hegedus and Heidi Carlone

E ducators have been thrown yet another new challenge: teaching engineering to their elementary school students. Challenge accepted! In this chapter, we highlight one teacher's journey with her fifth-grade students as they embarked on an engineering adventure together. Ms. Meriwether's experiences illustrate connections to the *Next Generation Science Standards* (*NGSS*) and emphasize opportunities engineering education affords students to think creatively and use 21st-century skills.

Ms. Meriwether was aware of the call to incorporate engineering practices into her science curriculum as part of the *NGSS*, but how would she get started? Whether driven by bravery or gumption, she headed forward full-steam and participated in a professional development session over the summer for training on the Engineering is Elementary (EiE) curriculum developed by the Museum of Science in Boston (2013). The EiE curriculum consists of 20 units, each highlighting a different field of engineering and science content, aligned with the standards. EiE was developed to provide project-based engineering challenges for students in grades

1–5. Ms. Meriwether chose the solar oven unit based on green engineering as her focus because of the close alignment with her district's fifth-grade standards and the fourth-grade national standards (see Connecting to the *Next Generation Science Standards*, p. 218, for solar energy connections). Energy is an important disciplinary core idea that spirals through the standards and grade levels. A "green" engineer develops and applies engineering solutions to environmental problems with a focus on minimizing the depletion of natural resources and reducing environmental impact.

Completing the EiE training was just one piece of the puzzle. Next, she was confronted with the reality of presenting the field of engineering and engineering practices to her students as a competent "expert." To her surprise, the journey was a relatively smooth and enlightening educational adventure. We describe Ms. Meriwether and her students' engineering journey here, showcasing students' experiences with creative problem solving along with practical strategies for implementing engineering practices into the daily curriculum.

We chronicle Ms. Meriwether's journey in four phases, corresponding to the structural arrangement of all EiE units (see Table 25.1). The four-phase EiE unit framework aligns nicely with the three core ideas of engineering design in the *NGSS*: (1) ETS1.A: Defining and Delimiting Engineering Problems; (2) ETS1.B: Developing Possible Solutions; and (3) ETS1.C: Optimizing the Design Solution (NGSS Lead States 2013). For teachers who do not have access to this curriculum, resources exist for developing solar ovens using everyday materials, engaging students in problem-based learning, and understanding the plight of people in developing countries by learning about their stories (see Internet Resources, p. 219).

TABLE 25.1

Overview of EiE Unit Structure

Part 1	An Engineering Story (set the context, encourage reflection by asking a series of questions, reinforce literacy skills)
Part 2	A Broader View of an Engineering Field (introduce the field of focus [green engineering—solar unit], discuss engineers' hands-on work in this field and the technologies produced)
Part 3	How Scientific Data Informs Engineering (link science, math, and engineering; collect and analyze scientific data to be referred to during the design phase)
Part 4	The Engineering Design Challenge (design, create, and improve solutions)

Defining and Delimiting Engineering Problems

Parts 1 and 2 of the EiE solar unit closely paralleled the first core idea of Defining and Delimiting Problems in the context of engineering. These parts of the unit provided rich opportunities for making meaningful connections to the social studies, science, literacy, and mathematics curriculum. What did this look like in practice?

Part 1: The Story

Because of predictable time limitations during the instructional day, Ms. Meriwether used her English language arts (ELA) time (an hour in the morning) to read the EiE story: *Lerato Cooks Up a Plan: A Green Engineering Story* (Museum of Science, Boston 2008) to students. Ms. Meriwether began with a summary of the story, providing visuals from the text on a projector. She initiated class discussions about the story's setting, context, genre, and characters. In the story, Lerato, a young girl in Botswana, is in charge of the family dinner, including the time-consuming and resource-intensive task of collecting firewood for cooking their food. Tsoane, her sister's friend who is studying green engineering at a university, gives Lerato a solar oven that does not require wood for fuel, but it does not work well at first. Lerato improves the design with knowledge of thermal insulators and conductors, though her precise solution is left intentionally vague. The problem of the story, how to build a well-insulated solar oven with reduced impact on the environment, set the context for the design challenge students would tackle later in the unit.

The book prompted questions about science and engineering issues (e.g., how the use of fires for heating can negatively impact the environment). Potential environmental impacts—aspects of green engineering—were discussed in the story and among the class members. One student commented that green engineers are people who "go out and use natural resources to solve problems in the world." Another student added that the word *environment* or *tikologo* includes "nature and our surroundings." Students discussed environmental impacts such as air pollution, soil erosion, and deforestation as issues that green engineers must consider to help preserve the environment. At this point, Ms. Meriwether initiated an activity to examine the use of paper (a naturally derived resource commonly used by students in the classroom).

Part 2: Introducing Green Engineering

In this session, youth learned about one strategy (assessments of natural resources) that green engineers use to define and delimit problems. The session began with practice in divergent thinking strategies. Students worked in groups of three or four to brainstorm as many uses for paper as they could in two minutes. The students became quite competitive, in a hurry to fill their lists with many ideas in the allotted time. Each group brainstormed 14–30 ideas, including toilet paper, name tags, school notebooks, paper plates, tickets, tests, lunch cards, tissues, signs, and so on. Students estimated the quantity of classroom paper used weekly by performing calculations, thus making mathematical connections. The brainstorming session extended into another activity in which students sequenced cards that represented the steps in the life cycle of paper (see Figure 25.1). This emphasized the importance of reducing, reusing, and recycling. The class discussed life cycle assessments as part of green engineers' jobs. Life cycle assessments are engineers' way of understanding the life cycle of human-made products, to include the examination of resources required for development, environmental impact, and any possible improvements that can be made. Students would use life cycle assessments later in the unit.

Part 1 of the unit set the stage by introducing the central engineering problem (how to create a well-insulated solar oven with the least

FIGURE 25.1
Life Cycle of Paper Cards

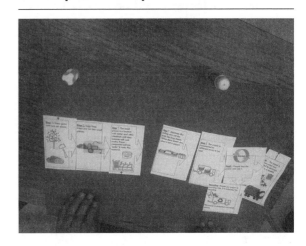

environmental impact) within the context of an engaging multicultural story. Part 2 allowed students to think like environmental engineers by conducting a life cycle assessment of paper to evaluate environmental impacts of using a natural resource. By understanding paper use in the classroom and the importance of reusing and recycling materials used on a daily basis, students were better able to grasp the need to protect natural resources (e.g., firewood) as they worked to develop alternative human-made technologies for cooking food. Defining and delimiting the problem (ETS1.A) in these ways provided a foundation for designing solutions for insulating a solar oven using materials with reduced environmental impact (parts 3 and 4).

Designing Solutions to Engineering Problems
Part 3: Testing Properties of Materials

The unit's next step involved testing different materials for their insulating properties. Students remained curious about the solar oven construction and offered their questions and ideas about materials they might use. One boy asked if the plastic wrap that goes on his sandwich was a good insulator. He pondered aloud how his lunch box was constructed by analyzing its insulating components. A girl commented that they would need to do something to the inside of the ovens to insulate them against heat loss. The students gasped with excitement when Ms. Meriwether presented the prototype oven (a modified shoe box) and let them know they would be making modifications on this design. Hands rushed up as students shared their ideas for how to design the boxes to yield the best insulation.

Students tested potential thermal insulators (craft foam, felt, newspaper, foil, plastic grocery bags, construction paper), changing the configuration of those materials (flat or shredded), developing ranking criteria (1–10 scale, with "1" meaning the best insulator and "10" meaning the worst), and evaluating materials based on their environmental impact (natural or human made, quantity needed, reusability, and recyclability). Cups lined with flat or shredded materials were placed in a large ice bath and secured in place to accommodate the 12 total testing stations (see Figure 25.2). The teacher secured 12 cups in advance to the bottom of the large plastic tub using a silicone adhesive. Students nested their testing cups inside the secured ones for stability. Ideally, there would be 12 student groups, 6 groups testing the flat and 6 groups testing the shredded insulators. Before testing, students made predictions about how the materials might perform in the test by considering their physical properties. Students monitored the temperature change of their assigned cup every 30 seconds for three minutes. Ms. Meriwether reviewed lab safety procedures before testing and instructed students in the reading and safe use of the school's glass thermometers. Safety goggles are recommended when using glassware in the lab. The class shared results and conducted an environmental impact analysis (see Figure 25.3)

FIGURE 25.2

Testing Potential Thermal Insulators

to determine the optimal materials, based on their insulating properties, for use in part 4, which is the design phase of the project. In part 3, students collected and analyzed scientific data about the insulating properties of materials and, in the process, made important linkages between science, mathematics, and engineering.

Optimizing the Design Solution

Part 4: Engineering Design

Students used their knowledge of thermal properties of materials and environmental impact (from part 3) along with the iterative engineering design process (EDP) (i.e., Ask, Imagine, Plan, Create, and Improve), to design their solar ovens in pairs (see Figure 25.4).

FIGURE 25.3

Environmental Impact Analysis

Environmental Impact Analysis: Class Data   A   B

Directions: Write each of the materials you will analyze in the boxes under "Material." Then, answer the questions to help you determine the environmental impact of each material.

		Material			
	Plastic	Foam	Newspaper	Aluminum foil	Felt
TYPE OF MATERIAL Is the material natural or is it processed?	Processed	Processed	Processed	Processed	Processed
REDUCE How much material do you need to use?	A lot	A little	Some	A lot	A little
REUSE Has the material been used before for another purpose?	varies	varies	varies	varies	varies
RECYCLE Can you recycle the material when you are finished with it?	Yes	NO	Yes	Yes	NO

(Waste Management)

EiE: Designing Solar Ovens
© Museum of Science, Boston
Duplication Permitted   3-10   Lesson 3: What's Hot & What's Not

FIGURE 25.4

Engineering Design Process

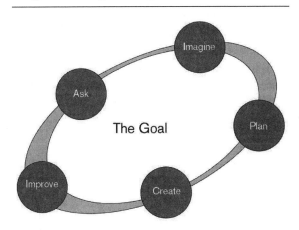

The Goal

Imagine · Ask · Plan · Create · Improve

Ms. Meriwether directed students to *imagine* or brainstorm some ideas for their design. They considered the advantages and disadvantages of using certain materials, the configuration of those materials, and how many "units" they would need. One unit of flat material was equal to one 8.5 × 11 sheet (e.g., felt or foam) or one cup of shredded material (e.g., newspaper or loose cotton balls). In doing so, they thought about the insulating properties of the materials in addition to their environmental impacts. The planning stage required labeled schematic diagrams for the distribution and placement of materials. Students used the Impact Scoring Sheet (see Figure 25.5, p. 216) to determine the trade-offs of using certain materials. For example, using natural materials resulted in a lower (desired) overall score versus the higher-scoring processed materials. Once constructed (created) in groups of two, students took ovens outside into a sunny location on the school grounds and began recording the change in temperature (with reflective lids open to direct the Sun's rays) at five-minute intervals for 30 minutes total (see Figure 25.6, p. 216). Students had proper sun protection to avoid overexposure when outdoors. At the end of this period, they moved the ovens to the shade and recorded the temperature change every minute for 10 minutes. Ms. Meriwether managed a control box (with no insulation) for comparison.

During testing, students scrambled around their boxes, shouting out temperature changes: "Mine is at 100 degrees!" "Ours is at 124!" Another female student was already thinking about how to improve her oven, exclaiming, "I know exactly what to do!" The session was not without its challenges. One boy stepped on and broke his thermometer in the excitement. Digital thermometers are recommended to avoid breakage and safety concerns, if school funds are available. Another group read the Celsius

FIGURE 25.5

Impact Scoring Sheet

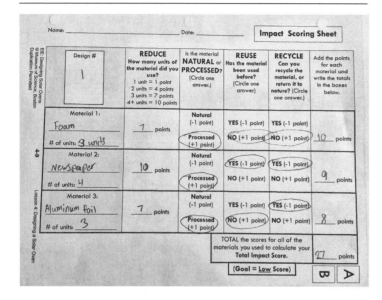

FIGURE 25.6

Recording Temperature of Solar Ovens in Sun and Shade

markings instead of the prescribed Fahrenheit readings, confusing their findings. All in all, however, the high level of student engagement and ideal afternoon weather conditions made the session a roaring success.

The next day, Ms. Meriwether conducted a class debrief of the temperature changes, and students calculated their total scores. After evaluating the oven designs that lost the least heat, students improved their designs and considered how to lower their environmental impact scores, retesting outside in a subsequent session.

What Did We Learn?

The *Now You're Cooking: Designing Solar Ovens* EiE unit aligns neatly within the three *NGSS* core ideas of engineering (ETS1.A, ETS1.B, and ETS1.C) and the *NGSS* energy performance expectation 4-PS3-4 (NGSS Lead States 2013). Additionally, the unit challenged students to think creatively and to experience productive moments of failure.

Thinking Creatively

Through the EDP, students experienced what it means to think creatively. First, students created multiple designs that presented more than one solution to a problem. Collaboration was an important element in this process. Sawyer (2012) noted that groups tend to be more creative than individuals and that cognitive diversity and group composition are critical elements in organizing creative groups. Second, students' divergent, flexible thinking emerged through brainstorming sessions about uses of paper and ways to insulate their solar ovens. Third, students developed novel ways to use ordinary materials in parts 3 and 4 of the unit. The design and improve phases prompted students' creative solutions, with opportunities to face a problem from different vantage points (see Figure 25.7).

FIGURE 25.7

How to Find Creative Solutions

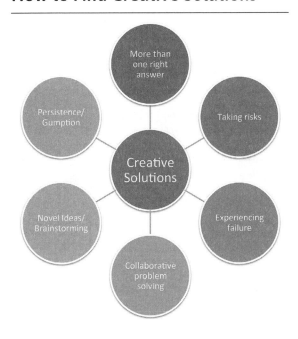

Productive Moments of Failure

Encouraging multiple solutions to a problem—that is, moving away from the traditional model of one right answer—can present challenges to students who are not used to experiencing failure. Throughout the unit, students repeatedly engaged with the problem and persisted in finding a viable solution. Ms. Meriwether emphasized this point with students during the improve phase of the design process. She let students know that her original design during a professional development session with EiE was not successful until she did some improvements. In this way, she shared her previous "failure" with students. Ms. Meriwether asked students if they would "feel like a failure if they did not experience the desired results?" One boy responded, "You don't want to feel like a failure." Ms. Meriwether reiterated to the boy and the class that "*you* are not a failure" if the design is not initially successful. She guided their attention back to the work of green engineers who learn from their previous designs flaws and use that information to improve future designs. The improve phase of the EDP provided students with new ways of approaching and solving problems, moving away from traditional right-or-wrong thinking.

Conclusion

Although implementing engineering practices may sound daunting, Ms. Meriwether showed that it is not an impossible proposition. In this case, literacy, social studies, mathematics, science, and engineering were integrated by making use of a multicultural story to engage students and set the stage for an engineering-based problem. Students could act as scientists by conducting controlled experiments to explore the properties of materials. They collected, analyzed, and reported their findings in a collaborative, scientifically oriented manner to gain knowledge about how to solve an authentic problem. Finally, the EDP (Ask, Imagine, Plan, Create, and Improve) allowed students to think creatively, experience productive failure, and optimize their design solutions in a real-world fashion. You can, too; take the challenge!

Connecting to the *Next Generation Science Standards*

The materials, lessons, and activities outlined in this chapter are just one step toward reaching the performance expectations listed below. Additional supporting materials, lessons, and activities will be required.

4-PS3 Energy www.nextgenscience.org/4ps3-energy	Connections to Classroom Activity
Performance Expectation	
4-PS3-4: Apply scientific ideas to design, test, and refine a device that converts energy from one form to another	Tested materials for their insulating properties and used that knowledge to build a solar oven with the goal of reduced impact on the environment
Science and Engineering Practices	
Asking Questions and Defining Problems	Participated in class discussions and questioning to define the problem in the story
Planning and Carrying Out Investigations	Conducted tests on materials to determine their insulating properties
Analyzing and Interpreting Data	Analyzed data to determine the best insulating materials to use in their solar ovens
Constructing Explanations and Designing Solutions	Constructed, tested, and improved solar oven designs
Disciplinary Core Ideas	
PS3.B: Conservation of Energy and Energy Transfer • Energy is present whenever there are moving objects, sound, light, or heat.	Used solar energy to heat food in an insulated solar oven
ETS1.A: Defining and Delimiting Engineering Problems • Possible solutions to a problem are limited by available materials and resources (constraints). The success of a designed solution is determined by considering the desired features of a solution (criteria). Different proposals for solutions can be compared on the basis of how well each one meets the specified criteria for success or how well each takes the constraints into account.	Generated lists in small groups or whole class discussion regarding the criteria and constraints of the problem presented Participated in class discussion using questions about the story to encourage reflection Exposed to the specific field of engineering indicated in the problem by exploring the hands-on work done by engineers in this field and the technologies produced
Crosscutting Concepts	
Energy and Matter	Tested and used various materials to enhance the transfer of energy
Influence of Science, Engineering and Technology on Society and the Natural World	Improved the solar ovens

Source: NGSS Lead States 2013.

References

Museum of Science, Boston. 2008. *Lerato cooks up a plan: A green engineering story*. Boston, MA: Engineering is Elementary

Museum of Science, Boston. 2013. *Now you're cooking: Designing solar ovens*. Boston, MA: Engineering is Elementary.

NGSS Lead States. 2013. *Next Generation Science Standards: For states, by states*. Washington, DC: National Academies Press. *www.nextgenscience.org/ next-generation-science-standards.*

Sawyer, R. K. 2012. *Explaining creativity: The science of human innovation*. 2nd ed. New York: Oxford University Press.

Internet Resources

Journey to Forever: Solar cooker resources

http://journeytoforever.org/sc_link.html

Library of Congress: Selected resources on solar ovens and solar cooking

www.loc.gov/rr/scitech/SciRefGuides/solarovens.html

Solar Cookers International: Solar cooking in the classroom

www.solarcookers.org/involved/teachers-and-students

TeachEngineering: Creating a solar oven

www.teachengineering.org/view_activity. php?url=collection/duk_/activities/duk_solaroven_ tech_act/duk_solaroven_tech_act.xml

26

Smashing Milk Cartons

Third-Grade Students Solve a Real-World Problem Using the Engineering Design Process, Collaborative Group Work, and Integrated STEM Education

By Debra Monson and Deborah Besser

Interdisciplinary projects that allow students to solve real-world problems can be highly engaging and motivating. These projects give students opportunities to apply, integrate, and expand their STEM knowledge. This chapter describes the third-grade portion of a project that highlights the use of integrated STEM education with an engineering focus. Students were engaged in the engineering design process (EDP) while the teacher provided guidance, information, and coaching. Although the teacher had an idea in mind, she chose to let students come up with the project idea through a series of informational class sessions.

The teacher facilitating this project collaborated with a kindergarten teacher and a sixth-grade teacher to create this project, which featured different problems in each of the three grades. The kindergarten class spent time learning about engineering, composting, and building individual composters. The third-grade class built machines to crush milk cartons using the engineering design cycle as a model for instruction. Meanwhile, the sixth-grade class worked on building a storage unit for the crushed milk cartons.

Engineering Design

Engineering design requires creative problem solving while considering constraints such as time, money, materials, and ease of use. The third-grade students used a design process that included defining a problem; generating alternatives; developing a workable design; analyzing that design; creating, testing, and improving; and coming up with a final product. As with many real engineering tasks, the design process often is not a linear route to a solution, but an iterative process. The model shown in Figure 26.1 (p. 222) of an EDP cycle illustrates what third-grade students were engaged in during this project.

Define the Problem

The third-grade teacher began the unit by helping her students gather information to fully define a problem and understand the constraints. Students were shown videos about composting and the use of landfills as a way to open discussion on a problem they may be able to solve (see Resources). The teacher explained that they

FIGURE 26.1

Engineering Design Process

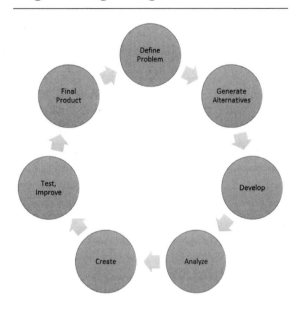

would be studying recycling and what garbage can do to the environment. She told students they would be watching videos to help them understand the effect that humans have on the environment.

After watching the videos about composting and landfills, students brainstormed ways they could reduce the impact of the garbage they produce at home or school. The teacher was thrilled that a student suggested exploring ways to recycle or compost the milk cartons they use at lunch, as that was the problem she intended to have them solve.

In a summer course focused on engineering education, the kindergarten, third-grade, and sixth-grade teachers at this school created a unit plan that they thought could be implemented the following year, based on the EDP. The summer course instructors encouraged the participants to begin with a real-world problem. This milk carton problem was the real-world problem the teachers thought the students would embrace.

They chose it with the hope that all three grades would be able to learn part or all of the engineering design cycle through the exploration of a solution. By introducing students to the ideas of recycling and composting and by showing them the excess garbage produced by the milk cartons, the third-grade teacher was hopeful that the students would realize this was a problem they could solve. Alternately, she could have introduced the problem and done essentially the same activity through a more direct approach, but her goal was for the problem to come from the students. Therefore, she took the chance that the students would come to the same conclusion as she had.

The teacher brought the students to the cafeteria to see all the garbage generated from lunch. She invited Bob, the custodian, into the classroom to share how the milk cartons, especially those with milk still in them, added to his workload. Students in third grade initially defined the problem as reducing the amount of garbage that Bob had to haul but, along with that solution, students realized they could compost or recycle the milk cartons. To solve this problem, students needed to find a way to empty and store the milk cartons for recycling or composting. This introduced a new problem because the space necessary to store the milk cartons would be great. Students then defined the real problem as developing a way to crush the milk cartons to reduce space until the cartons could be composted or recycled. The third graders focused on emptying and crushing the milk cartons, while the sixth-grade class worked on creating a storage container. The kindergarteners focused on the impact of composting.

Students collected data on the number of milk cartons used per day during both snack and lunch time and graphed this data over time (see Figure 26.2). They researched how long it takes various lunch items to decompose and graphed this data (see Figure 26.3). This information was

FIGURE 26.2

Milk Cartons Used Over Time

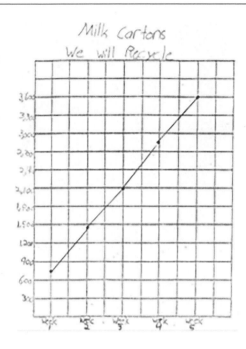

FIGURE 26.3

Decomposition Time for Various Lunch Items

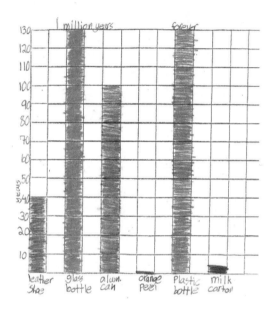

added to the personal story of the custodian to help the students clearly define the problem and begin to create a solution.

Generate Alternatives and Develop and Analyze Solutions

In keeping with the EDP, prior to any building, students individually brainstormed solutions and used graph paper to draw potential solutions and generate alternatives. Students were provided some materials to think about as they began developing their solutions: soup cans, a potato masher, 6 in. × 3 in. wooden blocks, PVC valves, other cylindrical shapes, and rectangular prisms. Students brought in a bucket and some string to add to the materials available.

Materials that were brought in needed to be student-friendly and have a very low (or no) risk of injury to the students. Beyond material constraints, it was identified that ALL students, from kindergarten to eighth grade, needed to be able to safely use the machine to crush their milk cartons. No models were created at this point; this stage was just a brainstorming phase to generate alternatives and develop solutions.

Students were then placed into heterogeneous groups of three or four, with both boys and girls in each group. They shared individual solutions, brainstormed beyond those solutions, and synthesized an optimal solution. The teacher remained a facilitator and let the students decide which solutions were reasonable, without trying to lead them down a set path. They analyzed their own solution, as a group, to choose one to create.

FIGURE 26.4

The Winning Design

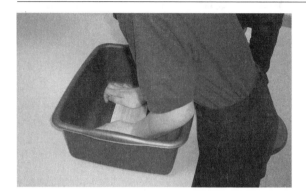 

Create, Test, and Improve

Students moved into the final prototyping phase in their small groups. Prototypes were created and then tested—first by third graders and then by kindergarteners—with the requirements that the machine could easily crush the milk carton and kindergarteners were able to use the machine. The teacher was careful to monitor the use of the prototypes during creation and testing to eliminate any safety concerns. Goggles were worn to avoid injury. Critical to the learning process was the teacher-facilitated discussion on possible improvements and steps needed for redesign. Questions students were asked included the following: How easy is it to use? How effective is it? How quick is it? Are there improvements that can be made? Students then made improvements and prepared to present to the class.

Final Product

The final product of each group was then presented to the class and put up for a vote on the best choice to fit the needs of the problem. To be the most successful project, the carton crusher needed to be effective, safe, quick, and inexpensive. The winning design (by class vote) ended up being the simplest: a pail to pour milk in, followed by a block of wood to crush the milk carton in a pan to prevent splashing (see Figure 26.4).

This solution, however, created a new problem, as so often happens in the real world. The kindergarten class was unable to crush their own cartons with the machine. So, how did the students solve this new problem? They agreed that the older students would simply help them crush their cartons while the classroom teachers supervised the crushing. This "non-engineering" solution served as an important lesson in the need for the "4Cs": communication, collaboration, critical thinking, and creativity.

Connections

The teacher in this classroom led a student-driven engineering design challenge that integrated science, engineering, and mathematics as well as other content areas and skills. Students were actively engaged in groups, sketched possible problem solutions, voted on designs, and wrote reflections on the process. Students were interviewed after the final product had been chosen and asked, "Will you please tell me about your engineering project?" Students were articulate when describing the process they went through and clearly stated environmental and practical reasons for doing this project.

An unexpected element that surfaced during the interviews was the students' compassion for the custodian and the awareness of the amount of garbage they were generating from the milk cartons. A student explained:

"We tested how much garbage we throw away a day and we got somewhere around two hundred pounds, I think. And Bob says it's really heavy. Each day we recycle milk cartons, we'll reduce about 15 pounds on each bag, so it'll be lighter for Bob."

Implementing Your Own Challenge

This was the students' first introduction to engineering and integrated STEM. Every student interviewed (about half the class) responded positively to the challenge, stated they would like to do another challenge, explained how actively engaged they were, and agreed that having failures was perfectly acceptable. This project helped further the students' understanding of an engineering design cycle, created opportunities for collaboration, and encouraged them to solve a problem they felt was important.

As teachers plan for their own real-world engineering challenge, they should consider some important things. First, can students help define the problem? If they can, what information do they need to be able to do so? Next, what are the constraints? Will you allow them to use unlimited materials, choose what they use, or attach a cost

to each material used? How can you highlight engineering design throughout the challenge? For this challenge, success was measured through the completion of a solution, but teachers can consider other formative assessments along the way to ensure active student engagement throughout the project. It is important to include check-in points throughout the activity to monitor progress of the individuals as well as the groups. The instructor assessed students' participation in activities through informal observation based on participation, used a KWL chart (Ogle 1986) to assess student understanding at various points in the process (what do you Know, what do you Want to know, and what did you Learn?), and used classroom discussion, including student answers to verbal questions, as a way to check progress and understanding. The teacher collected their graphs, observed group discussion, and carefully monitored student engagement. In addition to informal assessments, a more formal rubric could be used to give students feedback based on the EDP used (see NSTA Connection, p. 228).

One piece of this integrated work that could be improved was taking time to explicitly point out the connections between different content areas. When students in the interviews were asked if they did mathematics, most responded that they did not, even though they mentioned the graphing and calculations they had to do. The teacher agreed that next time she would like to spend time discussing the different content necessary to complete a problem like this. Making connections is a key component as we look to encourage all students to learn STEM concepts in the classroom.

Connecting to the *Next Generation Science Standards*

The materials, lessons, and activities outlined in this chapter are just one step toward reaching the performance expectations listed below. Additional supporting materials, lessons, and activities will be required.

2-PS1 Matter and Its Interactions *www.nextgenscience.org/2ps1-matter-interactions* **K-2-ETS1 Engineering Design** *www.nextgenscience.org/k-2ets1-engineering-design*	**Connections to Classroom Activity**
Performance Expectations	
2-PS1-2: Analyze data obtained from testing different materials to determine which materials have the properties that are best suited for an intended purpose	Built models to observe and test the ability of the crushing mechanism and materials to provide a solution that works for all students
K-2-ETS1-1: Ask questions, make observations, and gather information about a situation people want to change to define a simple problem that can be solved through the development of a new or improved object or tool	Watched videos on recycling, graphed data on composting, and met with the custodian to identify the problem of too much garbage with a solution of composting or recycling milk cartons and creating a crusher to reduce used space
K-2-ETS1-2: Develop a simple sketch, drawing, or physical model to illustrate how the shape of an object helps it function as needed to solve a given problem	Drew sketches of designs for the milk carton crushers and subsequently built a working model
K-2-ETS-3: Analyze data from tests of two objects designed to solve the same problem to compare the strengths and weaknesses of how each performs	Presented designs and, as a class, determined which design best fit the needs
Science and Engineering Practices	
Planning and Carrying Out Investigations	Planned simple investigations of materials and mechanisms available to solve the problem. Collaboratively planned, conducted, and investigated possible solutions and collected data
Analyzing and Interpreting Data	Analyzed data collected from testing to determine if the materials and mechanism solved the crushing problem
Constructing Explanations and Designing Solutions	Constructed explanations of which materials and mechanisms best solved the problem
Engaging in Argument From Evidence	Engaged in argument from evidence concerning success of concept

2-PS1 Matter and Its Interactions www.nextgenscience.org/2ps1-matter-interactions **K-2-ETS1 Engineering Design** www.nextgenscience.org/k-2ets1-engineering-design	**Connections to Classroom Activity**
Disciplinary Core Ideas	
PS1.A: Structure and Properties of Matter • Different properties are suited to different purposes. • A great variety of objects can be built up from a small set of pieces.	Evaluated the materials and mechanisms available to solve the problem Used a variety of materials to create the carton crusher
ETS1.A: Defining and Delimiting Engineering Problems • A situation that people want to change or create can be approached as a problem to be solved through engineering. • Asking questions, making observations, and gathering information are helpful in thinking about problems. • Before beginning to design a solution, it is important to clearly understand the problem.	Used the milk carton problem as an engineering design problem Watched videos and collected data before meeting with the custodian and learning about the problem with the trash Identified the problem as a class and set out parameters of the solution
ETS1.B: Developing Possible Solutions • Designs can be conveyed through sketches, drawings, or physical models. These representations are useful in communicating ideas for a problem's solutions to other people.	Sketched, shared, and created models of their solutions
ETS1.C: Optimizing the Design Solution • Because there is always more than one possible solution to a problem, it is useful to compare and test designs.	Shared solutions as a class and voted to determine the best solution based on the initial information and parameters
Crosscutting Concept	
Cause and Effect	Used models to test crushing performance and, in turn, gathered evidence to support or refute claims

Source: NGSS Lead States 2013.

References

NGSS Lead States. 2013. *Next Generation Science Standards: For states, by states.* Washington, DC: National Academies Press. *www.nextgenscience.org/next-generation-science-standards.*

Ogle, D. M. 1986. KWL: A teaching model that develops active reading of expository text. *The Reading Teacher* 39 (6): 564–570.

Resources

Gibbon, G. 1996. *Recycle! A handbook for kids.* New York: Little Brown Books for Young Readers.

Stover, J. 1989. *If everybody did.* Greenville, SC: JourneyForth.

Woolhouse, C. The Three Rs: Reduce, Reuse, Recycle. YouTube. *www.youtube.com/watch?v=wtoeZ9Nkeqk.*

NSTA Connection

Visit *www.nsta.org/sc1507* for the rubric.

27

Creating a Prosthetic Hand

3-D Printers Innovate and Inspire a Maker Movement

By Kristin Leigh Cook, Sarah B. Bush, and Richard Cox

The power of three-dimensional (3-D) printing technology has grown exponentially in just the past few years—people around the world are using 3-D printers to prepare food, create tailored clothing, build cars and homes, and advance the medical field in ways that never seemed possible (Martinez and Stager 2013). Even in classrooms across our nation, 3-D printers have become increasingly common because of their affordability and ease of use. Given the focus the *Next Generation Science Standards* (*NGSS*; NGSS Lead States 2013) and the *Standards for Technology Literacy* (ITEA 2007) place on engineering and design, 3D printers are enabling students to create tangible design solutions and inspire models of applied science, technology, engineering, arts, and mathematics (STEAM).

Teachers have recently begun to capitalize on the benefits of using 3-D printers to facilitate makerspaces—spaces where people can gather to create something or learn how to—in elementary classrooms. Allowing young students to create and innovate their own design solutions to real-world problems emulates how real science and engineering is done. In this chapter, we explore how one teacher employed 3-D printing technology with fourth-grade students in a STEAM lab to design and create a prosthetic hand. Mr. Smith's (pseudonym) vision for the STEAM lab, which he conceptualizes as a makerspace where learners of all abilities use multiple disciplines to solve problems, is to inspire a love of learning in students and to create a space for authentic inquiry about their world.

Designing a Solution

The project-based unit resulted from an identified need in the community. A student at a school in the district who was born without a hand was having difficulty logging onto the computers at school and needed help. Specifically, she needed the ability to press the Control + Alt + Delete keys at the same time—which are on opposite sides of the keyboard. Although at a different school, Mr. Smith was approached by the girl's teacher and parents who knew about his STEAM lab and resources. Beyond having access to a 3-D printer, the STEAM lab was a classroom known for inquiry-based learning about real-world problems. Mr. Smith decided this would be an authentic opportunity to engage students in the design and engineering process called for in the *NGSS*. Although the girl and her school remained anonymous, Mr. Smith's students readily accepted the challenge of designing, building, and printing a prosthetic hand and arm to help her with school-related tasks.

Students attended Mr. Smith's class as a special area five days a week for 50 minutes during a six-week period in which they undertook engineering projects as a way to integrate STEAM content. Although Mr. Smith conducted this unit in a STEAM lab, this project could be embedded in a traditional elementary classroom because it incorporates many content area standards in the core subjects of science, mathematics, and technology. Similarly, if teachers do not have access to a 3-D printer but still want to incorporate the research and design aspects of this unit in their classrooms, they can locate the nearest 3-D printer to their school through the online digital design site, Tinkercad (see Internet Resources, p. 236). Although Mr. Smith conducted this project over the course of six weeks because of the structure of the STEAM lab, teachers could conduct similar projects in a shorter time frame given their specific students and classroom structures.

Week 1: Building Empathy and Defining a Purpose

Because it was important to begin this project by understanding the potential for meaningful impact, Mr. Smith presented the problem to his students and took care to build empathy for the girl who needed their help. In doing so, he asked his fourth graders to consider what it might be like to have just one arm instead of two. It was important to set a tone of respect, so Mr. Smith emphasized the importance of equal access and briefly reviewed the Americans with Disabilities Act (see Internet Resources, p. 236).

To explore equal access, students inventoried the school grounds—moving around the school and attempting to perform simple tasks such as using the restroom, washing their hands, opening doors, and logging onto computers with only the use of one arm. Through this guided exploration, students became aware of how difficult simple tasks are with only one arm and understood the seriousness of the project on which they were about to embark. They began questioning what solutions the school might offer to help students with physical disabilities (e.g., installing automatic doors and motion-activated faucets, lowering bookshelves in the library). Students completed an Accessibility Analysis and Evaluation form (see NSTA Connection, p. 236), which asked them to identify areas of concern throughout the school, and submitted their results and recommendations to the principal. Setting the stage with this purpose-building activity inspired awareness of the scope of the problem while building excitement and a sense of purpose to which students could stay connected and motivated by throughout the six-week project.

Week 2: Conducting Guided Research

Students were then separated into self-selected teams (five different design teams of four or five students each). Within each team, students first used online resources to research prosthetic hands and hand-arm anatomy as they brainstormed, sketched, and designed independently for about 30 minutes. The students then collaborated to develop one team design using the most promising ideas from each student's independent work, which necessitated their articulation of why their design was chosen and why it should be included in the final group design.

In selecting features, students were prompted to consider functionality, usefulness, and feasibility with regard to building. Teams began conducting research and documenting their findings on artificial body parts, the skeletal system, prosthetics, and inventions on a large poster board. Sources for research included classroom books such as Engineering is Elementary biomedical engineering unit "Erik's Unexpected Twist: Designing Knee Braces" (see Internet Resources, p. 236). Students used Google Safesearch to peruse ideas for developing a prototype online. At the end of the week, teams developed a schematic drawing on poster-sized paper based on their group vision for their prosthetic (Figure 27.1).

FIGURE 27.1

Team Schematic of Researched Elements for Design

Week 3: Creating a Blueprint

Engineers and designers always have a plan. Students learn that while it's fun to tinker, create, and explore on the fly, the level of seriousness of this project and its implications demanded planning as much as possible, including making a blueprint or visual aid to guide construction and evaluation. Students from each team were asked to create a blueprint for their design on Tinkercad, an online site for creating digital designs that are ready to be 3-D printed into physical objects

FIGURE 27.2

Sample 3-D Blueprint Created on Tinkercad

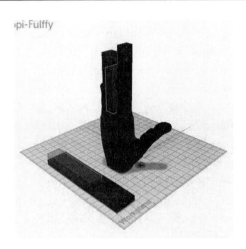

(see Internet Resources, p. 236). All students created a 3-D model on Tinkercad (Figure 27.2) and instructional prompts included the following:

- How can we make something both functional and aesthetically pleasing?

- How can we activate our schema and use ideas (from nature, from our daily lives) to inform our designs?

- How can we translate a two-dimensional idea into three-dimensional space?

Tinkercad is free, so teachers can set up one account and share the username and password with the students (Tinkercad requires anyone over the age of 13 to use an e-mail to create an account). Teachers can sign up students under their classroom account or receive parental permission for each student to have their own account. A classroom account allows teachers to keep track of all student designs and offer formative suggestions.

Week 4: Building the Prototype

Given a total classroom budget of $30–$50 for materials, teams developed a supply list of desired resources that weren't already readily available in the STEAM lab. Once all supply lists and budgets were approved, Mr. Smith visited a home improvement store website to order these specific amounts of materials. Because students knew the measurements for the desired prosthetic hand, they determined the appropriate sizing for arm design and attachments of the prosthetic to the body. Once the materials arrived, each team built a prototype (Figure 27.3). During the prototype phase, students tinkered and explored. Because many had never used small tools such as door hinges, PVC pipe, and foam insulation before, it was essential to allow time for exploration of the building materials. Safety considerations included a discussion of proper handling of materials and use of gloves and goggles. During this phase, students evaluated their prototype daily with a rubric in terms of the development and application of new and previous knowledge toward the larger design goal (Table 27.1).

FIGURE 27.3

Prototypes for a Prosthetic Hand

Week 5: Engaging in Public Relations

Next, teams presented their prototype to the class and other stakeholders to make arguments for and justify their design. The principal, the

TABLE 27.1

Team Schematic of Researched Elements for Design

Criteria	Proficient	Developing	Novice	Score
Development: Does my work show how my understanding of this topic changed over time?	5: I edited this work several times. This work shows my personal understanding in a clear way.	3: I edited this work once, and it shows my basic understanding of this topic.	1: I did not edit this work, and it does not show my understanding of this topic.	
Application: Does this piece make natural connections across ideas to create a work with an original message?	5: My work shows that I can make connections with two or more ideas to create something original.	3: My work shows understanding of the topic, but the connection to other ideas is not evident.	1: I did not make any connections to other ideas in my work, and it does not show originality.	

district technology integration specialist, other teachers, and classmates served as the audience for these presentations. The goal of the presentations was for each team to contribute ideas to an overall design for the next phase in which the teams merged into one company. Audience members voted on components of each design from each group that they would like to see included in the final product. This enabled there to be one final design that would be printed in three dimensions. With Mr. Smith as CEO, the company was reorganized into different departments and students were allowed to choose their group of interest from the following departments: challenge managers, rapid reporter or public relations, chief architects, and testing coordinator or quality control.

Week 6: Finalizing the Design

The challenge managers department began creating the final design in Tinkercad (Figure 27.4), attending to elements needed for mobility such as moveable and articulated fingers as well as durability to press keyboard buttons. The rapid reporter or public relations department tweeted and blogged about the experience on the class web page, added photos and descriptions of the project to the classroom web page, and were responsible for knowing all details of the project, such as the current stage of production or problems being trouble-shooted. Rapid reporters also managed the classroom DIY.org account (see Internet Resources, p. 236) in which students earn classroom badges for design-related innovations, identifying an opportunity to earn the Rapid Prototyper badge. The chief architects department coordinated building for the team (putting the prosthetic together), made design decisions during building (i.e., how pieces should connect, what pieces were necessary), ensured the team was

building safely at all times, and requested additional building help from classmates if needed or guidance from the company CEO. The testing coordinator or quality control department coordinated the tests needed to check success for the company, created a standard by which the final product would be judged (i.e., functionality, durability, aesthetics), and requested additional time if needed.

FIGURE 27.4

3-D Printed Prosthetic Hand

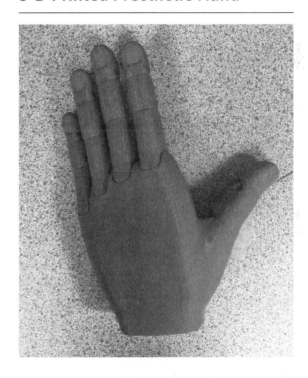

The actual 3-D printing of the prosthetic was completed during school hours; however, printing did not warrant the use of instructional time. Printing is time consuming. Teachers should focus students' time in the classroom on design and redesign, visualization, and simulation. Printing can be completed overnight, but if done during class time there are some safety precautions (Figure 27.5, p. 234). The students' creation, with its flexible fingers and attachable design, was useable by the student, enabling her to simultaneously touch the sides of the keyboard with ease.

FIGURE 27.5

Safety Precautions for 3-D Printing

- Print nozzle and heated plate is extremely hot. The separation tool to remove items from printer is sharp. Teachers should caution students to look but not touch and should wear heat-resistant gloves when removing items from printer.

- The melted plastic in the printer produces fumes. Printers should be used in a well-ventilated area.

STEAM Town Hall

In an effort to stay connected to the overall purpose of the project and provide formative feedback, Mr. Smith held a daily STEAM town hall meeting at the beginning of each day. During this time, students reflected on the previous class session's work, established learning objectives and guidelines, reviewed essential questions, and were introduced to a college or career focus that related to the project (e.g., biomedical engineering). STEAM town hall set the tone daily for rigorous and purposeful engineering work. The structure of each class session was essentially 10 minutes for the meeting, 40 minutes of teamwork, and 10 minutes for cleanup and debriefing using exit slips (Figure 27.6).

FIGURE 27.6

Exit Slip

Team Einstein — 4
How have today's experiences inspired me?
What was my biggest success today?
What was my biggest failure today?
How would I reteach what I learned?

How is what was created...	Great makers practice good habits!
Beautiful? Thoughtful? Personally meaningful? Sophisticated? Shareable? Moving? Enduring?	BE PROACTIVE BEGIN WITH THE END IN MIND PUT FIRST THINGS FIRST THINK WIN-WIN SEEK FIRST TO UNDERSTAND THEN TO BE UNDERSTOOD SYNERGIZE SHARPEN THE SAW

Conclusion

This project-based unit provided a memorable learning experience for Mr. Smith's class of devoted and motivated fourth graders. Moreover, this STEAM experience modeled interdisciplinary learning that is authentic, real-life, and which addresses a local need while simultaneously addressing standards found in the *NGSS* and the *Common Core State Standards* (NGAC and CCSSO 2010) for mathematics and standards for technology literacy. Students were engaged as a class, felt like an integral part of the important mission with which they were tasked, and looked forward to their time each day in the STEAM lab. We hope this chapter inspires other classroom teachers to seek timely and authentic learning opportunities for their students that integrate STEAM content areas and create lasting learning memories.

Connecting to the *Next Generation Science Standards*

The materials, lessons, and activities outlined in this chapter are just one step toward reaching the performance expectations listed below. Additional supporting materials, lessons, and activities will be required.

3-5-ETS1 Engineering Design *www.nextgenscience.org/3-5-ets1-1-engineering-design* **4-LS1 From Molecules to Organisms: Structures and Processes** *www.nextgenscience.org/4ls1-molecules-organisms-structures-processes*	**Connections to Classroom Activity**
Performance Expectations	
3-5-ETS1-1: Define a simple design problem reflecting a need or want that includes criteria for success and constraints on materials, time, or cost	Presented with a problem concerning a need for a prosthetic hand capable of accomplishing specific tasks; budget provided that restricts the materials available for development
3-5-ETS1-2: Generate and compare multiple possible solutions to a problem based on how well each is likely to meet the criteria and constraints of the project	Worked in teams to create designs that address the problem and then selected one for the creation of the product based on how successfully it functions
Science and Engineering Practice	
Constructing Explanations and Designing Solutions	Generated multiple solutions to the problem Explained how their design meets the criteria constraints of the design problem Determined a final product solution
Disciplinary Core Ideas	
ETS1.A: Defining and Delimiting Design Solutions • Possible solutions to a problem are limited by available materials and resources (constraints). The success of a designed solution is determined by considering the desired features of a solution (criteria). Different proposals for solutions can be compared on the basis of how well each one meets the specified criteria for success or how well each takes the constraints into account.	Worked in teams to create designs that address the problem and then selected one for the creation of the product based on how well each one meets the specified criteria for success
ETS1.B Developing Possible Solutions • Research on a problem should be carried out before beginning to design a solution.	Investigated problems confronted by those with one hand, how to solve problems, and tools involved

3-5-ETS1 Engineering Design www.nextgenscience.org/3-5-ets1-1-engineering-design 4-LS1 From Molecules to Organisms: Structures and Processes www.nextgenscience.org/4ls1-molecules-organisms-structures-processes	Connections to Classroom Activity
Disciplinary Core Ideas	
4-LS1.A: Structure and Function • Plants and animals have both internal and external structures that serve various functions in growth, survival, behavior, and reproduction.	Researched the skeletal and muscular systems and the ways in which humans use their hands
Crosscutting Concepts	
Systems and System Models	Researched and designed a prosthetic body part that mirrors animal species' abilities to interact with their environment
Cause and Effect	Determined the best prosthetic design based on proficiency at completing routine tasks

Source: NGSS Lead States 2013.

References

International Technology Education Association (ITEA). 2007. *Standards for technological literacy: Content for the study of technology.* 3rd ed. Reston, VA: ITEA.

Martinez, S., and G. Stager. 2013. *Invent to learn: Making, tinkering, and engineering in the classroom.* Torrance, CA: Constructing Modern Knowledge Press.

National Governors Association Center for Best Practices and Council of Chief State School Officers (NGAC and CCSSO). 2010. *Common core state standards.* Washington, DC: NGAC and CCSSO.

NGSS Lead States. 2013. *Next Generation Science Standards: For states, by states.* Washington, DC: National Academies Press. *www.nextgenscience.org/next-generation-science-standards.*

Internet Resources

Americans with Disabilities Act *www2.ed.gov/about/offices/list/ocr/docs/hq9805.html*

Buck Institute for Education (BIE): Rubrics *http://bie.org/objects/cat/rubrics*

Do it Yourself (DIY) *https://diy.org*

EiE: No Bones About It: Designing Knee Braces *www.eie.org/eie-curriculum/curriculum-units/no-bones-about-it-designing-knee-braces*

Tinkercad *www.tinkercad.com*

Tinkercad Instructions *www.3dvinci.net/PDFs/GettingStartedInTinkercad.pdf*

NSTA Connection

Download Accessibility Analysis and Evaluation form at www.nsta.org/SC1512.

Gliding Into Understanding

A Paper Airplane Investigation Highlights Scientific and Engineering Practices

By Patrick Brown

The crowd erupted into a roar as the paper airplane's journey came to an end. "773 cm!" A student shouted as her group recorded the data. As other students jotted down notes about ideas they were curious about, another group had timers at hand for the next round of flights, and others were busily tabulating data using calculators. Behind their work, these fifth-grade students were exploring plane flight and how scientists investigate the natural world.

A rich science learning experience not only captures students' attention but also motivates them to investigate and solve problems and investigate how scientists carry out their work. I have had success teaching students the nature of science by engaging them in tried-and-true investigation of paper airplanes (Silvis 2010). The paper airplane investigation relates to students' personal experiences and interests and cultivates enthusiasm by allowing students to develop ideas about forces, motion, and interactions, supporting *Next Generation Science Standards (NGSS)* 3-PS2 Motion and Stability: Forces and Interactions (NGSS Lead States 2013). This focus on a third-grade disciplinary core idea

and performance expectation was based on pre-assessments of student knowledge. They held incomplete views of the forces acting on objects in motion.

As a strategy for heightening interest and engagement, teachers can use the 5E instructional model (Engage, Explore, Explain, Elaborate, and Evaluate). The 5E model is beneficial for students because it highlights the importance of exploratory opportunities before explanations (Bybee 1997). Students that master this content can "Plan and conduct an investigation to provide evidence of the effects of balanced and unbalanced forces on the motion of an object" (3-PS2-1). As students progress through activities, they use the eight *NGSS* scientific and engineering practices (especially Asking Questions and Defining Problems, Planning and Carrying Out Investigations, and Engaging in Argument From Evidence) and the crosscutting concept Patterns (NGSS Lead States 2013).

Finally, the lesson includes many chances to bridge *Common Core State Standards (CCSS)* for Literacy and Mathematical Practices (NGAC and CCSSO 2010) with science content. As a result of the combined 5E and *NGSS* focus on the paper airplane activity, students develop deeper conceptual understanding and a greater ability to perform controlled science experiments for the remainder of the school year.

Engage Phase
One 60-minute class period

The engagement phase is a chance to elicit students' prior knowledge and experiences and motivate them to learn science. When students entered the classroom, I challenged them to work in groups of three to create a paper airplane using a half sheet of paper. Students predicted how far in centimeters their paper airplane would fly and described the forces acting on a paper airplane in flight according to experiences in earlier grades. I emphasized that although students may want to immediately test their ideas, they should not throw their paper planes at this point. Students' predictions lent insight to their knowledge of metric units. Some students thought the paper airplanes would fly only 10 cm, while others thought the planes would fly 100 m. In addition, their ideas about forces acting on a paper airplane in flight revealed their knowledge of force and motion. Students' predictions are formative assessments and inform the design of subsequent activities.

Explore Phase
Two 60-minute class periods

Once students have had a chance to work together, I facilitated discussions to help students explore paper airplanes. First, I asked students to think of an investigative question given the available materials and directions they had received so far. Students noticed there were three main types of paper airplanes identifiable by the nose design: "dart," "normal," and "glider"

FIGURE 28.1

Three Types of Paper Airplanes

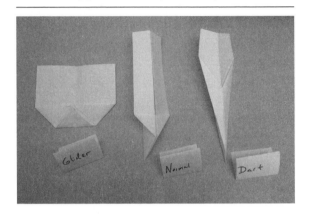

(see Figure 28.1). By honing in on the materials they could test and what they had already done (made a paper airplane), students decided as a class to investigate the question, "Does the style of paper airplane (dart vs. normal vs. glider) influence how far it travels?"

Students were now eager to test their planes; however, it is beneficial to discuss with students some science content behind airplane flight to develop a basic theoretical understanding and ensure that tests are reasonably accurate. The discussions about science content and experimental procedures are a good way to (formatively) assess students' content knowledge and thoughts about experimental design.

We started by constructing lists of forces acting on a paper airplane in flight (Figure 28.2) (*NGSS* performance expectation: 5-PS2-1). I called on volunteers to share ideas. Most students realized a paper airplane has an applied force (from the thrower). Some students mentioned that the force of gravity pulls the airplane down toward the ground. Few students identified air resistance as a force and no students mentioned lift forces exerted by the airplane on the surrounding air.

The next part of the class was focused on what we called "fair test" issues to ensure that when

FIGURE 28.2

Forces Acting on Paper Airplanes

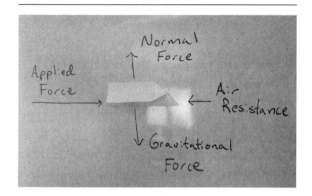

students conducted the experiment they used similar data collection procedures and followed safety practices. Generating a list of fair test issues and a scientific procedure addresses the *CCSS* for literacy because students have to design "steps for a technical procedure" (CCSS.ELA-LITERACY.RI.3.3; NGAC and CCSSO 2010). As a class, students generated the fair test issues and safety guidelines listed in Figure 28.3. I let students take turns presenting ideas as I wrote their thoughts on large poster paper for everyone to see. The end result of the discussions was that students developed ownership of learning because they played an active role in generating the exploration (science and engineering practice: Planning and Carrying Out Investigations; NGSSS Lead States 2013).

Once students had developed the investigation, they explored the relationship between a paper airplane's style and the distance it travels. Students worked in groups of three that included students with each type of paper airplane. Each group of students constructed the paper airplanes. Each student was assigned a role, either "thrower," "measurer," or "data recorder," and the students rotated roles each trial. To help manage the classroom and ensure there was enough space, we tested the paper airplanes in the

FIGURE 28.3

Fair Test Issues and Safety Guidelines for Conducting Investigations Using Paper Airplanes

1. One partner marks where the plane lands.

2. The distance flown is the point where the plane hits the ground.

3. Redo flights when plane's flight is disrupted (hits a table or wall).

4. Throw plane from approximately 2.5 m high.

5. Conduct three trials.

6. Do not disrupt others' planes.

7. Wear protective eyewear when conducting experiments.

8. When not performing data collection trials, sit by the wall of the gymnasium out of way of testing corridors.

gymnasium. For safety reasons, we used masking tape to mark off testing corridors so many different groups could test their paper airplanes at the same time. Students who were testing wore protective eyewear. When students were not conducting trials, they cheered on their classmates and stayed out of the testing area by sitting against the gymnasium wall.

Explain Phase
Two 60-minute class periods

The first portion of the concept explanation phase was dedicated to compiling data and making scientific claims based on evidence. I put a data chart on the board with spaces to list paper airplane style, trials, totals, and means. Students

TABLE 28.1

The Relationship Between Paper Airplane Style and Distance Traveled

Style of paper airplane	Distance Traveled (cm)					
	Prediction (cm)	Trial 1	Trial 2	Trial 3	Total	Mean
Dart	100	750	727	773	2,250	750.00
Glider	50	115	107	105	327	109.00
Normal	90	435	423	400	1,258	419.33

copied the table onto their own paper and worked with their groups to fill the chart with their data (see Table 28.1). They then worked together to analyze the data, looking for patterns between the mean distance traveled and style of paper airplane. I assessed students' abilities to calculate totals and means (including units) and provided feedback so they could accurately perform those calculations in subsequent activities.

Next, I asked questions to help students make sense of the data. For instance, I asked students to look at the highest mean numbers and lowest mean numbers to determine whether there was a relationship between distance traveled and style of paper airplane. This addressed the *CCSS* mathematical practice of using quantitative data to reason abstractly (CCSS.MATH. PRACTICE.MP2; NGAC and CCSSO 2010). Students quickly noticed a relationship and that the dart-style paper airplane consistently flew farther than the normal or glider paper airplanes.

Once students made a claim based on evidence, they engaged in argumentative discourse concerning why certain paper airplanes fly farther than others. I encouraged students to present ideas, ask questions, and justify claims based on evidence when they were discussing paper airplane flight (CCSS.ELA-LITERACY.SL3.1; NGAC and CCSSO

2010). Students generated a flurry of ideas. One student said, "Darts flew the farthest because they are more aerodynamic." You may find that students use the term *aerodynamic* without understanding how the shape of an airplane and an air resistance force influence paper airplane flight. I questioned students to think about the differences among the paper airplanes using force diagrams. Students realized the force of air resistance is greater for glide and normal paper airplanes, which have greater frontal surface area, than for dart-style paper airplanes. I also introduced students to the idea that the size of the arrow can be used to illustrate the magnitude of a force. Students had little trouble comprehending that a larger throwing force is depicted by a longer arrow. Finally, students discussed whether an object that is not moving has forces acting on it. Students realized that when objects are not in motion the sum of the forces is zero—not that there is no force acting on the object. Students demonstrated this understanding with their explanation that gravity is always acting on an object; therefore, other forces must also be acting that are equal in size and opposite in direction (standard 5-PS2; NGSS Lead States 2013). The explanation discussion engages students in generating reasons for their thinking—a skill students will use the remainder of the year when conducting more open-ended inquiries (science and engineering practice: Analyzing and Interpreting Data; NGSS Lead States 2013).

Once students have provided explanations for their investigations and discussions with fundamental science practices (science and engineering practices: Asking Questions and Defining Problems and Engaging in Argument Based on Evidence; NGSS Lead States 2013), I introduced

TABLE 28.2

Sample Conclusion Statements

Conclusion 1	Conclusion 2	Conclusion 3
Do different types of paper air planes affect how fast they are? The answer is yes. Darts are best and gliders are worst. That is what I found in my experiment.	If we try 3 different types of paper airplanes, gliders, darts, and normal. The type of paper airplane that was the fastest was the dart at 12 meters on average. The type of paper airplane that was the slowest was the normal at 10 meters on average. In the end, the best type of airplane was the dart.	In this lab we investigated whether the type of paper airplane affects speed. We found that the dart was went the farthest and that normal went the second farthest. In addition, we found that both the dart and normal go farther than a glider. We found that the glider went 4.26 meters, the normal went 5.67 meters, and the dart went 9.32 meters. In conclusion, the distance a paper airplane travels is dependent on the type of airplane.

formal science terminology in light of students' experiences. For example, I introduced the term *independent variable* (manipulated) to refer to the factor purposefully changed and the *dependent variable* (responding) to indicate the factor measured and reported as the results.

During the explanation phase, students also investigated scientific conclusions. Students were provided three different scientific conclusions and considered their attributes to derive a common set of guidelines for writing their own conclusion. For example, one sample conclusion statement had a misspelling, provided the research question, answered the investigative question, but did not offer data or evidence. The second conclusion had an incomplete sentence, provided a research question and data to support a claim, but made no scientific claim. Finally, the third conclusion provided the research question and data and states the relationship between the variables in the experiment (see Table 28.2). From the examination of different conclusion statements, students learned that a well-written

conclusion identifies the problem being investigated, generates a scientific claim based on data and evidence, and states the relationship between the variables in an experiment. Thus, students learned that a scientific conclusion is a specialized way of writing a summary of the most significant findings and relationship between variables of an experiment (CCSS. ELA-LITERACY.RIL3.3; NGAC and CCSSO 2010). As a result of the concept explanation phase, students derived knowledge of formal terminology and concepts through firsthand involvement with data, prior experiences, examples, and discussions with the teacher.

Elaborate Phase
Three 60-minute class periods

The aim of the elaborate phase is for students to apply new ideas in similar contexts. Students were challenged to work in their groups of three to design a "next step" investigation. The next step investigation required students to design a new and different scientific research question to explore paper airplanes. Students were instructed to change only one variable and to think about the factor they would measure and report on as investigation results.

Students were very excited about the elaborate phase and generated a flood of unique ideas. Some students wanted to test whether the size of the paper used to make paper airplanes influences the distance they travel. Others wanted to know whether the type of paper (construction, cardstock, computer, and lined paper) used to make paper airplanes influences the distance they travel. Several students tested whether changing mass by using paper clips affects airplane flight. A few students wanted to know whether paper airplanes travel farther outside or inside.

Once students decided on a research question as a group, they carried out their investigation in the gymnasium and wrote up lab reports, including the procedure followed, data tables and graphs, force diagrams (models for the forces acting on a paper airplane according to the data collected), and conclusions. This investigation was a great way for students to build and elaborate on their knowledge—an essential skill—of forces and motion (CCSS.ELA-LITERACY.W.3.7; NGAC and CCSSO 2010). I aimed to provide students investigative time and to ask probing questions to ensure students changed only one factor in their investigation (summative assessment).

Evaluation Phase
One 60-minute class period

Like scientists, students gave a brief, five-minute presentation that described their next step research question and main findings. The next step laboratory write-up and presentation were the culminating activities for the unit and the summative assessments of students' understanding of scientific practices. In addition, the lab write-up is

a way for students to "examine a topic and convey ideas, concepts, and information through the selection, organization, and analysis of relevant content" (NGAC and CCSSO 2010, p. 42).

I checked each component of students' projects. I graded students' data tables and graphs for accuracy and checked their diagrams to make sure their force diagrams include gravity and allied forces. In addition, I assessed students' written conclusions to make sure they provided the following: (a) a summary of their investigation; (b) data for the highs and lows; and (c) a statement describing the relationship between the independent and dependent variables in their investigation (see NSTA Connection, p. 244). Students enjoyed presenting their unique investigations and were proud of their work as scientists.

Conclusion

This type of investigation captures students' attention and unveils knowledge about the science process. Using the 5E instructional model (an exploration before explanation sequence) and the *NGSS* and *CCSS* as guides sets high standards for critical thinking, active participation in discussion, and knowledge generated through experiences with data and teacher-student interactions. The paper airplane investigation helped my students better conceptualize the parts of a scientific investigation when they compare the whole class paper airplane activity to their next step investigation. In addition, the paper airplane investigation helps students understand that only one variable can be manipulated in a controlled experiment to draw conclusive claims based on evidence. Students realized that they can change only the style of paper airplane in the first investigation and only one factor in their next step experiment to make valid scientific claims based on evidence (AAAS n.d). In many ways, the paper airplane investigation promotes long-term understanding of science because students actively engage in generating research questions, data collection techniques, and constructing claims based on evidence and data. Through the combination of hands-on, minds-on experiences provided by the paper airplane exploration, students learn essential scientific practices necessary when conducting investigations.

Connecting to the *Next Generation Science Standards*

The materials, lessons, and activities outlined in this chapter are just one step toward reaching the performance expectations listed below. Additional supporting materials, lessons, and activities will be required.

3-PS2 Motion and Stability: Forces and Interactions *www.nextgenscience.org/3ps2-motion-stability-forces-interactions* **3-5 ETS1-1 Engineering Design** *www.nextgenscience.org/3-5ets1-engineering-design*	**Connections to Classroom Activity**
Performance Expectations	
3-PS2-2: Make observations and/or measurements of an object's motion to provide evidence that a pattern can be used to predict future motion	Explored the relationship between paper airplane style and distance plane traveled to learn about the aerodynamics and flight distance, collected baseline data to investigate pattern, and investigated to test pattern in a new and different context
3-5-ETS-2: Generate and compare multiple possible solutions to a problem based on how well each is likely to meet the criteria and constraints of the problem	Worked in groups to create a paper airplane, predict how far plane will travel, and describe the forces acting on a paper airplane
Science and Engineering Practices	
Asking Questions and Defining Problems	Investigated whether the style of a paper airplane influences how far it traveled
Planning and Carrying Out Investigations	Developed a list of "fair test" issues to ensure testers use similar data collection and safety procedures
Engaging in Argument From Evidence	Explored different examples of conclusion statements to derive a set of attributes for a well-written conclusion that makes a scientific claim based on evidence
Disciplinary Core Ideas	
PS2.A: Force and Motion • Each force acts on one particular object and has both strength and a direction. An object at rest typically has multiple forces acting on it, but they add to give zero net force on the object. Forces that do not sum to zero can cause changes in the object's speed or direction of motion. • The patterns of an object's motion in various situations can be observed and measured; when that past motion exhibits a regular pattern, future motion can be predicted from it.	Investigated different forces influencing flight and learned that exerted forces can result in a sum of zero resulting in no motion and that forces can be represented with arrows indicating size and direction Developed conclusion statements including relationship between the style of the paper airplane and flight distance and supported claims with collected distance data

3-PS2 Motion and Stability: Forces and Interactions www.nextgenscience.org/3ps2-motion-stability-forces-interactions 3-5 ETS1-1 Engineering Design www.nextgenscience.org/3-5ets1-engineering-design	**Connections to Classroom Activity**
Disciplinary Core Ideas	
ETS1.A: Defining and Delimiting Engineering Problems • Possible solutions to a problem are limited by available materials and resources (constraints). The success of a designed solution is determined by considering the desired features of a solution (criteria). Different proposals for solutions can be compared on the basis of how well each one meets the specified criteria for success or how well each takes the constraints into account.	Designed a "next step" investigation to determine distance paper airplanes traveled changing only one variable based on available materials; wrote lab reports that included data tables and graphs, force diagrams, and conclusion statements
Crosscutting Concept	
Patterns	Worked together to analyze data, looking for patterns between the mean distance traveled and the style of paper airplanes

Source: NGSS Lead States 2013.

References

American Association for the Advancement of Science (AAAS) Project 2061. n.d. Pilot and field test data collected between 2006 and 2010. Unpublished raw data. *http://assessment.aaas.org.*

Bybee, R. W. 1997. *Achieving scientific literacy: From purposes to practices.* Portsmouth, NH: Heinemann Educational Books.

National Governors Association Center for Best Practices and Council of Chief State School Officers (NGAC and CCSSO). 2010. *Common core state standards.* Washington, DC: NGAC and CCSSO.

NGSS Lead States. 2013. *Next Generation Science Standards: For states, by states.* Washington, DC: National Academies Press. *www.nextgenscience.org/next-generation-science-standards.*

Silvis, K. 2010. Taking flight with an inquiry approach. In *Tried and true: Time-tested activities for middle school,* ed. I. Liftig, 1–5. Arlington, VA: NSTA Press.

NSTA Connection

Visit *www.nsta.org/SC1412* to download the representative scoring guide for students' presentation of data and conclusion statements.

A System of Systems

A STEM Investigation Project for Intermediate Students Has Real-Life Connections

By Barney Peterson

System: A group of interacting, interrelated, or interdependent elements forming a complex whole, or a functionally related group of elements.

"Hey, Mrs. P, what's that thing for?" "How come the ceiling looks like that?" "Why are there handles on those pipes in the hallway?" "How does that thing work?"

The first months in our new school were punctuated by dozens of student questions in the hallways, in the classrooms, in the cafeteria—everywhere we went. Our new school, besides being built to be environmentally friendly, is designed in a fascinating industrial-tech style that uses nontraditional finishes and exposes many of the normally hidden parts of a building to view. Because my students were so intrigued by things they noticed, the construction manager and I decided to capitalize on their curiosity and create an authentic learning opportunity that supports the growing emphasis on STEM education. Our goals for this unit were to help the students understand systems and system models in real-world contexts and to be able to distinguish the parts of systems and their functions. This relates to *Next Generation Science Standards* (*NGSS*) crosscutting concept System and System Models (NGSS Lead States 2013).

Students would also use the speaking, listening, technology, and teamwork skills we had developed to create a communications product for an audience beyond our classroom.

Learning to Use the Tools

Although our main project was still in the planning and development stage, I enlisted our school learning resource specialist to help students gain experience writing scripts and choosing visuals to share specific information (CCSS.ELA-LITERACY.CCRA.SL.5 connection: Speaking and Listening: "Make strategic use of digital media and visual displays of data to express information and enhance understanding of presentations"; NGAC and CCSSO 2010). Students were taught how to develop video presentations using two different platforms available at our school: PCs and iPads. Students worked with the learning resource specialist during weekly library periods and used daily literacy block time in the classroom to complete these projects. This experience would provide the skills and practice necessary for them to communicate what they learned in the project with a broader audience beyond the classroom.

Figuring Out the Systems

The students and I decided to take a systems approach to organizing our research. We were already familiar with the idea of systems from our yearlong units raising salmon in our classroom and studying watershed ecology to understand salmon habitats. Students had learned about body systems in salmon anatomy and how they support the survival of the whole animal. They had also studied watershed ecology from the viewpoint of river and stream systems, so they had a basic idea of how to search for and recognize the parts of a system. This background provided an opportunity to extend our understanding of systems to include physical science as well as life and Earth science. We posted "System: A group of interacting, interrelated, or interdependent elements forming a complex whole, or a functionally related group of elements" to guide us as we investigated what systems were present in our new school (*NGSS* crosscutting concept Systems and System Models: "A system can be described in terms of its components and their interactions"; NGSS Lead States 2013).

While the class discussed all the parts and functions we could think of that are normally required to make a building run, I created a list of their ideas on the whiteboard. As a group, the class walked around our school, enthusiastically discussing what we saw and taking pictures of everything that caught our eyes or piqued our curiosity. Students worked in teams of four, each team with one iPad. They captured pictures of switches, pipes, wires, holes, covers, color patterns, material

surfaces, and more. Back in the classroom, student teams gathered noisily around tables to examine their photos and delete duplicates, then I printed out paper copies so teams could sort and organize the photos into related sets, such as lighting, decoration, wiring, walls, and floors, based on what function we thought each might serve. Because our school was described by district officials as being "green," a team of three students researched "green construction" to help us understand that label. It was time to call in the experts!

Learning From the Experts

I enlisted the help of Darcy Walker, a district construction manager who was involved in building our school. Together, we planned how to organize this project. He worked with the class to narrow our ideas to a list of seven major systems: civil, data and communication, electrical and lighting, fire and safety, HVAC (heating, ventilation, and air circulation), landscape and design, and structural. From there, Mr. Walker and I sought speakers who could help the students understand the parts of each system and how they work, separately and together. He enlisted seven people involved in the design and construction of our building and scheduled their visits. They included both males and females from design and engineering fields to provide students with exposure to diverse career opportunities and to show them that accessibility to those careers is not limited by gender or ethnicity.

Earlier in the school year, guest speakers from the public utilities for water, gas, and electricity visited our class to help students learn about the importance of using sound construction methods and materials for efficiency and economy. They engaged students in hands-on experiences such as investigating how air flows through rooms, affecting heating, and the amount of energy used by various kinds of lightbulbs. Our new series of speakers built on students' understanding from those earlier experiences to help us discover environmentally friendly features of the building (3-5-ETS1: Engineering Design and crosscutting concept Influence of Engineering, Technology, and Science on Society and the Natural World; NGSS Lead States 2013). These included saving energy by using LED lighting, maximizing natural light with large classroom and clearstory windows, choosing a site drainage system with built-in filtration for surface water runoff, and developing ways of recovering heat from the heat recovery system to use in other ways in the building.

Speakers came equipped with samples of materials, construction drawings, and slide shows to help students understand special features of the new building. As exhibits were passed around and handled, students asked questions and responded to prompts from speakers to learn how the materials worked. Asking about levers on the big silver pipes overhead lead to discovering how dampers in the ductwork control the airflow through the ventilation system to constantly integrate fresh air from outside with heated air and provide a healthier atmosphere. When a student pointed out the silver "buttons" on classroom ceilings, we all learned about how the fire suppression sprinkler system is disguised with pop-off covers. While touching pieces of "rebar," students discovered that the "bumps" on the bar actually help to hold it solidly in place in the concrete foundations and walls of the building. Students expanded their understanding of "reduce, reuse, and recycle" to include energy as well as materials when they learned that, as our old school was demolished, it was actually carefully dismantled and materials were separated for recycling. The mechanical engineer showed how a particular device is used to recover heat from air that has already circulated through the school before it is pushed outside: This heat is used to help warm fresh air as it enters the system. A rewarding "aha!" moment

happened when a student said to me "Hey, that's like the radiator on our car, right?"

All of the speakers had prior experience working with students in a variety of formal and informal situations and needed very little coaching to prepare them for working with us. We encouraged them to bring visuals and hands-on examples whenever possible to tap into the variety of learning styles the students use.

Organizing a System of Systems

Meanwhile, students began re-sorting photos into the systems and posting them on our "A System of Systems" wall. This visual organizer was continually modified as our understanding of the systems and their subsystems grew (Figure 29.1, p. 248). Students began to question which systems were entirely separate and how the systems depended on each other to work. We discovered components that supported each other and were organized into systems that accomplish particular critical functions in the building. Students prepared for our speakers by discussing what we already knew about the systems and things we had observed during the building of the new school: the ground-shaping and forms built for footings and walls; trenches and plastic pipes through which wires ran; more metal, concrete, and bricks than wood used in walls; huge wooden beams in the roofs. Students recorded and discussed their questions as they sorted photos and listened to speakers. After the last of the individual system component videos were finished, a team of three students took on the task of organizing and explaining this in an overview video segment (see NSTA Connection, p. 253).

Because the new school was built within feet of our old facility, we had daily opportunities to observe the deconstruction and materials salvage of the old building as well as the ground-up

FIGURE 29.1

29

Section of the System of Systems Organizer Wall

building of the new school. The old school consisted of six hexagonal pods of classrooms and three semi-rectangular support units. Instead of hallways, there were covered walkways between buildings. Our architect used photos and drawings to help students work through how he designed a new, larger school to replace the old school on the same amount of property. We saw that the new school used the same area-footprint as the nine old buildings and walkways. By adding a second story, we had a larger school and still had room for playgrounds, an athletic field, and parking lots on the same property. The two-story building also made more effective use of a single HVAC system. Later, the civil and

structural engineers demonstrated how two-story construction required different materials and techniques to create a strong, safe, energy-efficient building. We used construction drawings to compare the new school to the old school. We learned to look at drawings and see three-dimensional figures as students discovered how math and reasoning skills are used in models to address the challenges involved in planning and building. For example, the way special foundations in the ground were required to support the weight of a large brick wall that stands alone as part of the roof design particularly intrigued the students. Doing so helped them learn about supporting greater weights with a large base to ensure stability and how the architect and structural engineers cooperated to design a workable system that complemented the overall design of the building. Students continued to refer back to the construction drawings throughout the nine months of this project. This helped them understand the spatial relationships between parts of systems and the working connections that kept all systems functioning. An example would be looking at wiring diagrams to see that the electrical system connected to almost everything else in the building.

Calling on Community Resources

Our first visitor was the lead architect for the project. He began with a wonderful analogy describing the work of planning and building a school as teamwork organized much like our local NFL team with a manager, coaches, and players who specialized in certain roles, but all of whom work together to win games. Before he ever got into the details of design, he helped the students gain new understandings of systems, both as the building and its parts and as the working groups of people who create it.

Speakers came two or three days apart: electrical, civil, structural, mechanical, safety, and design were all presented by professionals specific to each field, thus exposing the students to the science of building design and construction. As they listened to our speakers, students took notes and asked questions to grow their understanding of how so many parts are organized and coordinated into one huge functioning system. During his presentation, the structural engineer made a deliberate point of engaging the students in dialogue to discover how much they knew about the earthquake-prone area in which we live. He had prepared a special PowerPoint presentation to help the students understand how living in an earthquake-prone area

requires buildings to be stronger and more flexible (see NSTA Connection, p. 253). That, combined with a demonstration using a shaker table, helped students realize the teamwork between geoscientists and civil, structural, and geotechnical engineers that goes into making our public buildings safe. This relates to *NGSS* disciplinary core ideas ESS3.B: Natural Hazards and ETS1.A: Defining and Delimiting Engineering Problems (NGSS Lead States 2013).

Speakers also engaged students in conversations about career opportunities and the educational preparation required for taking advantage of those opportunities. This discussion proved to be one of the surprise rewards of the project as individual students began to pursue information about specific careers that interested them. I encouraged students to extend their relationship with our speakers, using them as a resource for additional information.

Sharing Our Information

After the speakers, it was time for students to reorganize and refine their information, linking it with specific images and then writing scripts for the videos they would create to explain our building and its systems to others. Students were assigned to groups of three or four with a mix of ability levels, genders, and language proficiencies. Students in this class often work in groups of this nature to support each other through joint research, peer editing, and English language acquisition. This communications phase was very labor intensive.

Mr. Walker took all students in small groups on behind-the-scenes tours of the school to explore areas not normally visible to the public. Because the active construction phase was finished and safety procedures were followed (such as staying in approved observation zones), additional safety equipment such as hardhats or safety glasses was not required. During the tours, students took photos and asked questions. The tours proved to be a valuable tactic, especially in expanding their knowledge of how buildings are designed with backup equipment against possible emergencies or failures.

Mr. Walker also took on the task of writing brief background narratives in plain talk, linking systems and subsystems together to help students organize their presentations. He and I then sat down with each system-team and discussed how they could write their scripts to ensure only accurate, important information was included. We concentrated on asking questions to reveal their understanding and any areas where they needed more clarification to report accurately. A priority for the project was having students share learning in their own language and style. Where necessary, we facilitated students finding or correcting information. This sometimes meant an extra visit to specific areas such as the boiler room or the elevator control room to clarify ideas about how things really work. Next, each team presented a draft script to other teams for peer review, focusing on clarity of information for communication with an audience that would know very little about how a building functions.

Once scripts were developed, teams found quiet spots to practice recording and playback for fluency and timing. They critiqued and perfected their audio together before recording the videos. Once live tracks and still photos were aligned with audio, finished videos were created and then uploaded to School Tube where they were assigned URLs which could be linked to QR tags (see Internet Resource, p. 253). We learned how to use the Tag Maker program to create a specific code for each video. The next step was to print the QR tags and post them in the school. The tags allow anyone in the school to use a Smartphone or tablet computer with a code-reader app to scan the QR tags and play the videos to learn about

our school as a "system of systems." The district facilities and planning department funded manufacture of high-quality, durable signs, to ensure that they would match school design.

Evaluating Our Work

Before any signs were posted, students had one final important step to complete. We had planned for evaluation right from the start of the project with students helping to establish the goals and criteria by which they would be evaluated at the end of the project. As a first step, we employed a two-color response form that students and I developed from the project plan. Each student used a yellow highlighter to color in boxes rating themselves on specific criteria, and then I used a blue highlighter to record my ratings. I met with students individually to discuss our ratings. When we agreed that student work was satisfactory (boxes showing green), the work was considered accepted. In areas where blue and yellow showed separately, we would discuss differences and plan how to resolve them. All areas were required to earn green coding before final videos were uploaded to get URLs and QR tags were made to put on signs.

Formal evaluation was developed to expand on the simple two-color response form. Where the response form was designed for approval before posting the teams' products, this formal rubric provided a tool for grading (see NSTA Connection, p. 253) (performance expectation 5-ESS3-1: "Obtain and combine information about ways individual communities use science ideas to protect the Earth's resources and environment"; NGSS Lead States 2013).

Students were individually evaluated on their engagement and participation as a team member and their personal contributions to the process. Each student had opportunities to demonstrate reaching objectives: recognizing systems as they were encountered and understanding the interdependency of their parts, conducting research using a variety of information resources; writing scripts; communicating verbally and in writing; and effectively using communications technology.

After videos were recorded, URLs were assigned and QR tags were created. We exposed our product to a test audience of students, teachers, and district employees. We evaluated feedback about how well the tour worked: Was it well organized? Were individual components and systems explained clearly? Were there areas we missed or important points we neglected? Did the tags work correctly to activate the videos? Were recordings clear? Did the audience understand the purpose of demonstrating how our school works as a "system of systems?" We made some adjustments in response to feedback we received.

Reflecting on the Journey

Throughout this project, students were encouraged to use their curiosity about unfamiliar things to extend their learning. They raised important questions, worked effectively in teams and with outside experts, and improved their communications skills. By engaging the students with Mr. Walker and speakers from the building team, we provided opportunities to learn about careers and the importance of post–high school education early in students' school experience. Commitment of time and resources by the facilities and planning department demonstrated how interdepartmental teamwork can provide powerful support for student learning. Because our students learned about how our school was designed and built and how it is supposed to function, they acquired a valuable sense of investment in our building. Producing the building tour to share with others was a big achievement for the students, but in the long run the greatest value and pleasure from this project came from

the investigation phase that actively engaged students with experts, ideas, and resources that were totally new to them.

Learning Results and Future Considerations

Students from the fourth-grade class who started this project were unable to complete it by the end of their fourth-grade school year, so 15 of the 24 original class members were recruited as fifth graders to reassemble in my classroom after school one day each week for three more months during the next school year to finish up their videos. In conclusion, they were given the opportunity to meet with me and Mr. Walker to reflect on the project. We included two of their current fifth-grade teachers who were new to our school. Students proudly gathered around an electrical and data systems closet and amazed their teachers with the depth and breadth of their knowledge about the identity and function of the sea of multicolored cables and connectors. In a round-robin conversation, they revealed what they had learned about how information is relayed through fiber optic cables, and they moved on to explain how the invisible-beam smoke detection system works. They had just begun to talk about the HVAC system when it was time for them to go home. The information they had retained and the depth of the questions they asked and answered was testimony to the effectiveness of this hands-on, active learning experience.

Our System of Systems project was not, according to the students, finished when the 21 short videos were completed and their signs were posted. A team of three fifth graders came back to my classroom twice each week for an additional two months in spring 2014 to create one more piece of the project: a longer video that examines all the systems and identifies one system that unifies them. This last video demonstrated just how much they had grown in their understanding of systems. They brought their project to a satisfying conclusion when they presented the video at the District's STEM fair at the end of the school year (see Internet Resource, p. 253).

For others considering a similar project, I would suggest having a very clear purpose from the start. Make sure students recognize that they will be communicating with both adults and children so they need to learn as much as possible about their topic before they try to share what they have learned. In a project this huge and so far outside the comfort zone of many children this age, working in small teams is critical. That will help ensure teamwork and teacher time are spread more evenly and visibly. Making use of outside speakers gave the research and instruction a special feeling that called for students to pay attention and learn as opportunities were presented. It is also critical to be very flexible and not set short deadlines on a project of this nature.

Connecting to the *Next Generation Science Standards*

The materials, lessons, and activities outlined in this chapter are just one step toward reaching the performance expectations listed below. Additional supporting materials, lessons, and activities will be required.

3-5-ETS1 Engineering Design www.nextgenscience.org/3-5ets1-engineering-design **4-ESS3 Earth and Human Activity** www.nextgenscience.org/4ess3-earth-human-activity	**Connections to Classroom Activity**
Performance Expectation	
3-5 ETS-1-1: Define a simple design problem reflecting a need or want that includes specified criteria for success and constraints on materials, time, and cost	Learned how teams of people worked together to define and solve problems to be met through design, engineering, and construction
Science and Engineering Practice	
Asking Questions and Defining Problems	Used photos and information from speakers to answer questions about structure, function, and purpose of specific components and learned how materials worked and how they worked together as systems to solve the defined problems
Disciplinary Core Ideas	
ETS1.A: Defining and Delimiting Engineering Problems • The success of a design solution is determined by considering the desired features of a solution.	Worked with the architect to understand how a new, larger school could be built on the same area-footprint as the old school and allow for playfields and parking on limited property
ETS1.B: Developing Possible Solutions • At whatever stage, communicating with peers about proposed solutions is an important part of the design process, and shared ideas can lead to improved designs.	Shared what they learned about the building's systems in by developing brief videos linked to QR tags that were posted around the building to define components of building systems and help others understand how systems work together

3-5-ETS1 Engineering Design *www.nextgenscience.org/3-5ets1-engineering-design* **4-ESS3 Earth and Human Activity** *www.nextgenscience.org/4ess3-earth-human-activity*	**Connections to Classroom Activity**
Disciplinary Core Ideas	
4-ESS3.B: Natural Hazards • A variety of hazards result from natural processes. Humans cannot eliminate the hazards but can take steps to reduce their impacts.	Worked with an engineer to learn about how geoscientists and civil, structural, and technical engineers work together to make a building stronger and more flexible for an earthquake-prone area
Crosscutting Concepts	
Influence of Engineering, Technology, and Science on Society and the Natural World	Researched green construction as a team and helped the class to understand how recycling materials from the old school's deconstruction and use of state-of-the-art materials and techniques in the new building made the new building resource and energy efficient
Systems and System Models	Worked with speakers to sort components and develop a list of seven major systems based on how the components worked together

Source: NGSS Lead States 2013.

References

National Governors Association Center for Best Practices and Council of Chief State School Officers (NGAC and CCSSO). 2010. *Common core state standards*. Washington, DC: NGAC and CCSSO.

NGSS Lead States. 2013. *Next Generation Science Standards: For states, by states*. Washington, DC: National Academies Press. *www.nextgenscience.org/next-generation-science-standards*.

Internet Resource

"A System of Systems Overview" video
www.schooltube.com/video/6907880ef74d463393fd

NSTA Connection

For a sample cooperative grading sheet, the systems template, and the formal rubric, visit **www.nsta.org/sc1501**.

Blasting Off With Engineering

Toy Testing Creatively Engages Fifth-Grade Students in Engineering Design

By Emily A. Dare, Gregory T. Childs, E. Ashley Cannaday, and Gillian H. Roehrig

3-2-1. Blast off! What better way to engage young students in physical science concepts than to have them engineer flying toy rockets? The *Next Generation Science Standards* (*NGSS*) and researchers alike (Brophy et al. 2008) advocate the integration of engineering into science classrooms because engineering provides the following:

- A *real-world context* for learning science and mathematics

- A context for developing *problem-solving skills*

- A context for the development of *communication skills and teamwork*

Engineering design is a central feature of the scientific and engineering practices in the *NGSS*, allowing students to solve problems using science and mathematics content in an engaging and meaningful context (NGSS Lead States 2013). Our toy rocket unit engaged fifth-grade students in an engineering design process that allowed them to integrate engineering, physical

Three students adjust their rocket to the correct launch criteria.

science, and mathematics content. The abilities of these students ranged from those with special needs, who were assisted by classroom aides, to students from a gifted and talented program. This unit tapped into students' creativity in the hopes that they would understand science and engineering as creative endeavors (NRC 2012). Online, we share alignment to the *NGSS* practices (see NSTA Connection, p. 261).

The unit was adapted from a NASA foam rocket activity (see Internet Resource, p. 261). We chose these foam rockets for their simple design and readily available, inexpensive parts: rubber bands, foam tubing, string, cable ties, card stock, and small masses from a classroom balance scale. Details regarding building the rockets and the launchers can be found on the NASA website. Over a 10-day period, students engaged in an engineering design process, collecting and using data from a series of science inquiry activities to investigate several factors that affect the rocket's flight (see NSTA Connection, p. 261, for a day-by-day breakdown). Students not only participated in mini-investigations, but also created their own rocket along with a final marketing product, such as a commercial advertisement or package, which they presented during a mini-conference. Students were moving toward meeting performance expectation 3-5ETS1-3: "Plan and carry out fair tests in which variables are controlled and failure points are considered to identify aspects of a model or prototype that can be improved" (NGSS Lead States 2013).

The Challenge

Setting an engineering lesson in a realistic, meaningful context is critical to engage students and motivate learning. We created this context on the first day when we presented students with a flying toy challenge memorandum from a toy company asking students to design a toy rocket based on an initial prototype (see NSTA Connection, p. 261). After discussing the term *prototype*, students were provided with three rocket design constraints:

1.   The flying toy's behavior must be predictable or easily adjustable.

2.   The materials must be common and inexpensive (provided by us).

3.   The toy must be safe for elementary school use.

By providing constraints along with a given challenge, students were able to move toward meeting the performance expectation 3-5ETS1-1: "Define a simple design problem reflecting a need or want that includes specified criteria for success and constraints on materials, time, or cost" (NGSS Lead States 2013). We also required students to create a product to promote their toy. Although each student created an individual rocket, students were allowed to work with a partner to create a marketing product. Their choices were an advertisement (radio or television), a letter to the company's research and development department, a package or container for the rocket, a questionnaire or survey to learn what rocket characteristics students prefer, and a chart that described their rocket's performance. As anticipated, many students chose to create packages because it was a fun, artlike project. We provided cardboard boxes and coloring utensils, though many students brought in items from home, such as shoeboxes or stickers.

After we introduced students to the design challenge, they needed to learn about the performance of the basic prototype to determine what modifications would be beneficial. The property that we tested was how far the prototype flew when launched. You will need a large space for launching, such as an empty cafeteria or gymnasium. Students wore safety goggles during the testing.

After we showed the students how to safely launch the rocket, we allowed them to take turns launching, taking care that other students were not in the rocket's path. This step served the purpose of collecting data from multiple trials, thus addressing the need to calculate an *average* flown distance. These trials also showed students that the rocket did not fly very consistently.

Students pull the rubber band back 30 cm and launch their toy rocket.

After this initial test, we asked students, "What factors affect how your rocket flies?" Whereas some student ideas were beyond our resources, such as adding propellers or a motor, most of their ideas aligned with testable variables: launch angle, rocket weight, rocket length, number of fins, and length of the rubber-band pull. This step allowed students to develop possible solutions as outlined in ETS1.B (NGSS Lead States 2013). We assigned students a variable to test in a small group, engaging them in planning and carrying out controlled experiments as outlined in the scientific and engineering practices (NGSS Lead States 2013). We provided students with prebuilt rockets and tables to record the data. Although this activity produced noise and chatter, students successfully completed multiple trials for each variable to calculate an average distance for their changed variable. Furthermore, this step allowed them to critically analyze the effects of the different variables.

Test and Share

As part of the community aspect of engineering, students shared their data with the class so that everyone benefited from the variable testing.

Students reported the "best" parameters to produce the longest flight distance (see NSTA Connection, p. 261, for a results table). Additional recommendations were made based on students' observations: (a) Although no mass meant the rocket flew farther, a 10 g mass might be more appropriate for indoor use because it did not fly as far; (b) increasing the mass led to more reliable results (i.e., the rocket tended to land in the same place during each trial); and (c) a shorter rubber-band pull for indoor use is best because the rocket did not fly as far. Students were able to synthesize their findings into interpretations (another practice), to provide us with the following relationships, with all other variables controlled:

- The more mass the rocket has, the more force required to launch the rocket.

- The more force applied to the rocket, the greater distance the rocket flies.

- The rocket travels the greatest distance launched at a 45° angle.

Students used this information to create their own toy rockets. We gave students two class periods to work on planning, building, and testing. Before building, students were required to develop a written plan or drawing, including a materials list. We asked each student to place an "order" so that we could create individual bags of materials between classes, alleviating chaos in distributing materials. Students were expected to do a final rocket launch, and we required them to report on the specs of their modified rocket. This approach gave students firsthand experience of "design-test-redesign" in our engineering design process, aligning with several of the *NGSS* scientific and engineering practices. We dedicated two additional class periods to working on final rocket design and accompanying marketing products; some students also chose to work at home.

30

For the final day, we invited parents and administrators to the students' "mini-conference" in the school cafeteria. The mini-conference structure allowed students to present multiple times, fielding questions about their final rocket designs and accompanying products (see NSTA Connection, p. 261). Groups who had created commercials gave live performances, their clever scripts generating laughter. We encouraged students to visit their peers' work, promoting the community aspect of science and engineering.

Assessment

During the mini-conference, we used a rubric to assess students' projects and presentations, which we had provided to students at the start of the unit (see NSTA Connection, p. 261); we asked students to self-assess using the same rubric. This rubric emphasized the importance of our engineering design *process*, as opposed to grading based on the "best" rocket (e.g., farthest distance or the prettiest). We wanted this unit to encourage students' creativity, which is an important feature to engage students and allow them to understand that creativity is integral to engineering (NRC 2012). Many students used humor in the creative process; for instance, a live-action commercial in which a television provided the viewer with a toy rocket through the screen. We also assessed students' presentations during the mini-conference, giving them some sense of what being an engineer or scientist is really like—that there is a social factor to both of these fields that allows one to engage with others and disseminate ideas to the general public. Here, students communicated their ideas, which is another scientific and engineering practice (NGSS Lead States 2013).

In addition to the assessments, we asked students to complete a satisfaction survey containing

Suggested Improvements

We learned several lessons throughout our first implementation regarding materials, such as the following:

- *New rubber bands*—older rubber bands have less elasticity, so that box of rubber bands from the back of a desk drawer is not a good option.

- *Non-finished string*—we used masonry line, which is slippery, causing rockets to fall apart. Something like twine or cotton string is likely to improve rocket structure.

- *Heavier rocket fins*—card stock is not very sturdy, thin corrugated cardboard might work better. Increasing the weight of the fins may also provide better stability in the rocket's flight.

Likert-scale (see NSTA Connection, p. 261) and free-response items. It was no surprise that surveys indicated students enjoyed this engaging and creative context. The free-response items indicated that many students' favorite part was, "Making the box or commercial," or simply, "It was fun." They reported the majority of their problems were in building a box or creating a commercial. When asked to report what they learned, many students addressed the fact that heavier rockets "created accuracy" and making a toy was more difficult than they originally thought.

It was interesting that many students reported that adding mass to their rocket would increase its stability and flight consistency, yet only a few of the final rocket designs included this feature. Creating a rocket with replicable results was one of the constraints voiced in the toy company memo, yet many students disregarded this, focusing instead on distance. We should have made it clear that when students were testing variables and measuring flown distance, they should also be taking notes about the consistency of the rocket's behavior.

Time constraints limited the amount of time spent directly talking about force and motion. When we repeat this unit in the future, we hope to better integrate these science discussions after the testing phases, asking students to point out patterns they see, thus leading to discussions of why these patterns exist, and introducing fundamental physics concepts. For example, although students intuitively understood that a longer rubber-band pull would launch the rocket farther, we should develop this idea using Newton's second law. Students' surveys reflect this since students did not feel as strongly about learning science as using creativity. Although our students enjoyed the unit and came away with some basic physics concepts,

we could improve this. Additional topics that could be included for older students include lift, thrust, drag, and gravity.

Another addition that comes to mind is using a trade book, such as *Leo Cockroach: Toy Tester* (O'Malley 2001), to lead into this activity. This book about a cockroach toy tester would be the perfect lead-in to get students thinking about the engineering design process and would add a strong literacy component. Despite having no trade book lead-in during our first implementation, this unit has laid a solid foundation that integrates concepts from science, engineering, and mathematics. It is clear from student reactions and survey responses that creativity is a key aspect to engage students in engineering design processes.

Connecting to the *Next Generation Science Standards*

The materials, lessons, and activities outlined in this chapter are just one step toward reaching the performance expectations listed below. Additional supporting materials, lessons, and activities will be required.

3-5-ETS1 Engineering Design www.nextgenscience.org/3-5ets1-engineering-design **3-PS2 Motion and Stability: Forces and Interactions** www.nextgenscience.org/3ps2-motion-stability-forces-interactions	**Connections to Classroom Activity**
Performance Expectations	
3-5-ETS1-1: Define a simple design problem reflecting a need or want that includes specified criteria for success and constraints on materials, time, or cost	Addressed the needs of the toy company client by meeting the memorandum's criteria, only using the limited materials available
3-5-ETS1-3: Plan and carry out fair tests in which variables are controlled and failure points are considered to identify aspects of a model or prototype that can be improved	Tested iterations of toy rocket models by changing one variable at a time (launch angle, rocket width, rocket length, number of fins, and length of rubber-band pull) and compared results to determine what features make the "best" model
3-PS2-2: Make observations and/or measurements of an object's motion to provide evidence that a pattern can be used to predict future motion	Made observations, collected data, and shared observations and data to collectively determine patterns in toy rocket flight performance (distance flown) across the class's complete data set

3-5-ETS1 Engineering Design www.nextgenscience.org/3-5ets1-engineering-design **3-PS2 Motion and Stability: Forces and Interactions** www.nextgenscience.org/3ps2-motion-stability-forces-interactions	**Connections to Classroom Activity**
Science and Engineering Practices	
Developing and Using Models	Modified the toy rocket prototype and tested each iteration to determine the most desirable features
Planning and Carry Out Investigations	Planned and conducted tests of the modified toy rockets, collecting data to share with their classmates
Analyzing and Interpreting Data	Collected data from testing different variables, shared data with their peers, and determined what features were desirable for their final product
Designing Solutions	Used the results of data analysis to design their own version of the toy rocket
Disciplinary Core Ideas	
ETS1.B: Developing Possible Solutions • Tests are often designed to identify failure points or difficulties, which suggest the elements of the design that need to be improved.	Participated in controlled variable tests to collect data about which features of the toy rocket would lead to success or failure
ETS1.C: Optimizing the Design Solution • Different solutions need to be tested in order to determine which of them best solves the problem, given the criteria and the constraints.	Shared the results of their controlled variable testing with peers to determine which features of the toy rockets would best address the client's needs
PS2.A: Forces and Motion • The patterns of an object's motion in various situations can be observed and measured; when that past motion exhibits a regular pattern, future motion can be predicted from it.	Compared data and observations with their classmates to determine the toy rocket prototype's behavior given different tested variables and used this information to make decisions about their own toy rocket models
Crosscutting Concepts	
Influence on Science, Engineering, and Technology on Society and the Natural World	Considered the needs of the client as well as the findings of variables testing to design and build a rocket that the client would find appealing
Patterns	Noticed patterns in the toy rocket's behavior to make informed decisions about their own models

Source: NGSS Lead States 2013.

References

Brophy, S., S. Klein, M. Portsmore, and C. Rogers. 2008. Advancing engineering education in P–12 classrooms. *Journal of Engineering Education* 97 (3): 369–387.

National Research Council (NRC). 2012. *A framework for K–12 science education: Practices, crosscutting concepts, and core ideas*. Washington, DC: National Academies Press.

NGSS Lead States. 2013. *Next Generation Science Standards: For states, by states*. Washington, DC: National Academies Press. *www.nextgenscience.org/next-generation-science-standards*.

O'Malley, K. 2001. *Leo Cockroach: Toy tester*. New York: Walker & Company.

Internet Resource

Rockets: Educator's Guide
www.nasa.gov/audience/foreducators/topnav/materials/listbytype/Rockets.html

---

NSTA Connection

View examples of stsudent work, the rubric, and more at *www.nsta/org/SC1411.*

SCAMPERing Into Engineering!

A "Snapshot of Science" Program Brings Science and Engineering Into the Library

By Jenny Sue Flannagan and Margaret Sawyer

"Whoa! Did you see that thing zoom off the shelf?" exclaimed one of the fourth-grade boys as his group watched him test his "airmobile" on the library counter. The excitement in their eyes was apparent as they stood by, waiting patiently to test their own cars next.

When our school division was asked to make more explicit connections to engineering, we wondered how we would fit it in. With the heavy emphasis on reading and math in elementary classrooms, adding science to an already crowded day seemed a daunting challenge. But in thinking outside the box a little and looking for natural ways to bridge engineering design projects with science units, we have found a way to not only make it work for us but also creatively engage students in what they love—science!

The lesson described here is part of what we like to call our "Snapshots of Science" program. These mini-lessons of science are taught once a week to all students in the school library. Yes—you read that right—in the school library! Over the past two years, we have been working to extend the experiences students have in their science classroom into the library. Each week, students experience a lesson that goes a little further than their

classroom experience. These snapshots are only 20 minutes in length and are tied to the *Common Core State Standards*. Although they are done in a short amount of time, we have found it is enough time to get students excited to learn more. By using the 5E approach to break our lesson unit into smaller parts, our students are able to focus on one concept or skill and build on what they learned each week while having time between our library lessons to discover and explore the material we are covering. These lessons provide students with an opportunity to do even more hands-on activities that are different from the ones done in their classroom. These snapshots also provide children with the freedom to explore their activities without having to take home their work each night for additional study.

The snapshot lesson SCAMPERing Into Engineering was designed around the fourth-grade curriculum unit Forces and Motion. Forces and Motion is a four-week science unit that engages students in the study of motion (see Figure 31.1 for what to know, understand, and do for the unit). Activities engaged students in designing simple investigations to test how mass is related to motion and how different surfaces affect friction, and students learned to identify and explain when objects exhibited potential and kinetic energy.

SCAMPER is a mnemonic that stands for

Substitute.

Combine.

Adapt.

Modify.

Put to another use.

Eliminate.

Reverse.

There are points in the science curriculum where engineering naturally lives—topics where there is a natural connection between science and the opportunity to integrate math and engineering. For instance, teaching simple machines and motion would easily provide opportunities to plan units during which students are actually able to design something that demonstrates the basic concepts and apply that material to an engineering activity. For us, the unit on motion was the perfect time to introduce students to the idea of engineering. We wanted our students to understand that engineers often take the research done by scientists and use it to solve a real-world problem. One real-world problem associated with cars and motion has to do with overcoming air resistance. Knowing that we could not have our students design a real car, we chose instead to engage them in designing a car when given some simple materials. To solve the problem, they would have to apply what they had learned and what they would learn about air

resistance and friction. This engineering problem also allowed us to introduce students to a simple strategy they could use in the design process.

Engage and Explore: Transferring Learning to New Situations

This lesson in the library occurred over four weeks. During the first week, students were given a simple yet challenging task—design a car that moves with a single breath of air using only the following items: four ring-shaped hard candies, two straws, two paper clips, scissors, tape, and a sheet of paper. They were given no instructions on *how* to build their vehicles and only told to design one they thought would work using air power. Students were also told there was no "right" or "wrong" answer; their goal was simply to design a wind-powered vehicle. Children who struggled to get started were given the same hint: think about things that move with air. The children at their table would then begin to offer ideas, such as kites, blimps, sailboats, hang gliders, and hot air balloons. Students were allowed to work as individuals or within small groups at their table. Most children chose to work alone, but the children at their tables often provided guidance when the student was faced with a dilemma.

Students were reminded of rules regarding scissor safety and responsible use of paper clips. Given that this activity did not have any rubber bands or other projectile pieces, we did not require all students to wear goggles, but they were available for students wishing to use them. When students tested their cars to see how far the vehicles would move using their breath, only one student was allowed at the counter at a time, while the other students remained at the group tables and observed from there. Constant teacher supervision was provided.

FIGURE 31.1

What to Know, Understand, and Do for the Unit

Know

- The position of an object can be described by locating it relative to another object or to the background.

- Tracing and measuring an object's position over time can describe its motion.

- Speed describes how fast an object is moving.

- Energy may exist in two states: kinetic or potential.

- Kinetic energy is the energy of motion.

- A force is any push or pull that causes an object to move, stop, or change speed or direction.

- The greater the force, the greater the change in motion will be. The more massive an object, the less effect a given force will have on the object.

- Friction is the resistance to motion created by two objects moving against each other. Friction creates heat.

- Unless acted on by a force, objects in motion tend to stay in motion and objects at rest remain at rest.

Understand

- The pattern of an object's motion in various situations can be observed and measured.

- Regular patterns of an object's motion can be used to predict future motion.

- Patterns of failure of a designed system can be used to improve design.

Do

- Describe the position of an object.

- Collect and display, in a table and line graph, time and position data for a moving object.

- Explain that speed is a measure of motion.

- Interpret data to determine if the speed of an object is increasing, decreasing, or remaining the same.

- Identify the forces that cause an object's motion.

- Describe the direction of an object's motion: up, down, forward, backward.

- Infer that objects have kinetic energy.

- Design an investigation to test the following hypothesis: "If the mass of an object increases, then the force needed to move it will increase."

- Design an investigation to determine the effect of friction on moving objects. Write a testable hypothesis and identify the dependent variable, the independent variable, and the constants. Conduct a fair test, collect and record the data, analyze the data, and report the results of the data.

Source: The Knows and Dos are from the 2010 Science Standards of Learning Curriculum Framework found at *www. pen.k12.va.us/testing/sol/standards_docs/science/index. shtml.*

As they built and tested their cars, students were instructed to take notes of what they observed the car doing. Students noted whether their cars moved, how their cars moved, and how far their cars moved along the counter. At the conclusion of this lesson, students were assessed using a 3-2-1 exit ticket. They shared three things they had observed about their car, two questions they still had about the cars or designs, and one thing they wanted to change during the next week's lesson. We felt this provided us with a glimpse into what students were observing and wondering.

Explain: Thinking Like an Engineer

During week 2, we introduced students to the concept of an engineer and explained to students what engineers do. This allowed for the perfect opportunity to introduce the SCAMPER brainstorming tool (see Figure 31.2). Developed by Bob Eberle (1984), SCAMPER is an acronym that stands for the words Substitute, Combine, Adapt, Modify, Put to Another Use, Eliminate, and Reverse. SCAMPER is essentially a cognitive strategy that serves as a scaffold (Palincsar and Brown 1984).

During instruction, we often want students to solve a problem or a task. Cognitive strategies are those tools that provide students with a structure for learning when a task cannot be completed through a series of steps. In moving students through the redesign phase of their car, there really wasn't a set of steps we could take them through. Because each car was designed differently, we had to find a tool that would work for all students and be applicable to their car designs. The SCAMPER brainstorming tool was the perfect cognitive strategy or scaffold (Rosenshine 1997) to use because it provided students with a set of questions they could "think" through when examining their own design.

To teach students the tool, we did not start with their car but rather with an object they were familiar with using each day—a toothbrush. This was done intentionally because we wanted students to focus on learning the tool without having to learn the tool and think about their new design. For example, in redesigning a better toothbrush, we asked students to think through the questions associated with Substitution. Could they substitute a different material in place of the material used in the original design that might make the toothbrush more effective or easier to use? Could they combine something in the design? As we explained to students, the questions were merely tools to help them think in different ways (Figure 31.2 includes the words and associated questions). Students worked in collaborative groups and worked with their teacher, Mrs. Sawyer, to go through each letter to rethink the design of their toothbrushes. When they were finished, students shared their ideas with the entire class.

Elaborate: Designing and Testing

Once we felt that students understood the strategy, we then had them go back to their original car designs. Using the SCAMPER method, students took what they did during the first week and redesigned their vehicles by answering certain questions and marking what they changed. Beginning with the vehicles themselves, the designs were as different as the children who built them. In fact, even with the children working with other students at their tables, no two "airmobiles" were alike. Many had a four-wheel design, similar to that of a car, but others resembled more of a sailboat or airplane. Designs included between two and eight wheels. Most used straws to build a sort of chassis for their vehicle, but a few used paper clips as their structural base. Shapes of the designs were varied; there were triangular,

FIGURE 31.2

SCAMPER

When to Use the Strategy

Use it when you want students to think creatively to change the design of something.

How to Use the Strategy

Consider an existing object, product, or service. Use the question stems to help you brainstorm things you could change or modify. When done, look at the various answers. Which ones could you try to see if they make a difference?

Substitute

- What materials or resources can you substitute or swap to improve the product?
- What other product or process could you use?
- What rules could you substitute?
- Can you use this product somewhere else, or as a substitute for something else?
- What will happen if you change your feelings or attitude toward this product?

Combine

- What would happen if you combined this product with another to create something new?
- What if you combined purposes or objectives?
- What could you combine to maximize the uses of this product?
- How could you combine talent and resources to create a new approach to this product?

Adapt

- How could you adapt or readjust this product to serve another purpose or use?
- What else is the product like?
- Who or what could you emulate to adapt this product?
- What else is like your product?
- What other context could you put your product into?
- What other products or ideas could you use for inspiration?

Modify

- How could you change the shape, look, or feel of your product?
- What could you add to modify this product?
- What could you emphasize or highlight to create more value?
- What element of this product could you strengthen to create something new?

Put to Another Use

- Can you use this product somewhere else, perhaps in another industry?
- Who else could use this product?
- How would this product behave differently in another setting?
- Could you recycle the waste from this product to make something new?

Eliminate

- How could you streamline or simplify this product?
- What features, parts, or rules could you eliminate?
- What could you understate or tone down?
- How could you make it smaller, faster, lighter, or more fun?
- What would happen if you took away part of this product?
- What would you have in its place?

Reverse

- What would happen if you reversed this process or sequenced things differently?
- What if you try to do the exact opposite of what you're trying to do now?
- What components could you substitute to change the order of this product?
- What roles could you reverse or swap?
- How could you reorganize this product?

Source: Glen 1997.

FIGURE 31.3

Various Cars Designed by Students

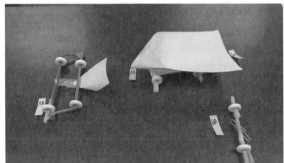

rectangular, and square designs. Some students made 3-D shapes like a rectangular prism or square pyramid. Still others built aspects of their vehicles from origami shapes (Figure 31.3 shows a variety of car designs).

Students used the prior knowledge of wind-powered vehicles as the basis for their projects. Many of the vehicles were built with a design that resembled a sailboat with wheels. Some featured wings similar to a biplane. Still others featured a kite-inspired design with cross-supports and a piece of paper stretched over the frame. In an effort to catch the wind and force their vehicle to move, many of the vehicles featured a sail (triangular or rectangular). Others included a parachute design that students believed would allow

the air to fill in the chute and cause the vehicle to move. Ultimately, they saw how the tiny things they altered on their vehicles led to some interesting and sometimes incredibly different results in distance and speed.

Evaluate: Lessons Learned

It was amazing to see how students worked through this design challenge. Students worked together, and in some cases, students who never wanted to work with anyone else were jumping in to provide support and ideas to their classmates. Waiting to test their designs did not matter; students were just as excited to see how their classmates' redesigns worked as they were to test their own cars. The students watched their classmates' airmobiles and paid attention to how each design worked. They rarely spoke, except when encouraging each other or in making note of how something responded to the air.

"Did you see how the car turned sideways? I don't think it was supposed to do that, but it really went far!" noted one of the boys. "I think I'm gonna try a rectangle sail if my triangle one doesn't work after my first run." The children were so interested in the designs that they couldn't wait to get back to their tables and see how to improve them!

Using the SCAMPER process, one student rebuilt her vehicle "like a hovercraft." The wheels, rather than being attached to roll, were attached to the bottom of straws to remain flat on the surface of the table. A sail-like device was then attached to allow the vehicle to move as the candy "wheels" glided along the smooth table, rather than rolling on an axel. The student said she did this because the tape she had originally used to attach the wheels to an axel "got in the way of it moving. The wheels couldn't roll well because they kept getting stuck. Then, a wheel would quit turning and my car would blow sideways and fall off the table." She found that, after reengineering her vehicle, her distance more than doubled.

Another girl in one of the fourth-grade classes was not immediately interested in participating, and instead made origami flowers with straws as the stems. Rather than redirecting her, we chose to let her continue folding her paper as her table mates worked feverishly to design their vehicles and see what happened. She had made four of the flowers and held them together in a bouquet, when someone noted to her that the straws came together like they were part of a car frame. She decided to see if she could indeed use her flowers to make a vehicle. She taped the four straws together at the end and then attached a wheel to

the point using a paper clip. After taping the flowers together, she discovered she had accidentally created a wind-powered vehicle with a car frame that resembled a Formula One design. Her vehicle not only went far but also straight. After her first trial, she was immediately hooked and went back to her table to see how she could improve her car to go even farther. She was so excited to discover her accidental success in engineering!

Students found unexpected successes as they designed and redesigned throughout the SCAMPER process. One child had made an origami "puff box," believing that as he blew on the side of the box with the hole, his breath would cause the box to inflate and fill up with air, thus catching his wind and moving along the table. What he discovered, however, was that blowing from a distance onto his box did not cause it to inflate. Instead, he found that his breath was caught on the un-inflated side of the box, causing his vehicle to move sideways. He used SCAMPER and bent some of the origami folds and flaps out, and as a result, he discovered his car was able to go farther in distance and straighter as it moved.

Assessing Student Thinking

Often during learning experiences such as the SCAMPER lesson, a formal assessment is not necessary. Our goal with the assessment tied to this lesson was to see the student's thinking and identify any questions they still had in their minds. To quickly assess, we like to use the strategy known as 3-2-1. Students reflected on the three things they learned through the design process; two things they still had questions about with regard to changing their design, and one thing they felt really good about during the experience. This assessment was done after week one and again after the introduction of the SCAMPER strategy. When the students came back to the library, we were able to address the questions provided to us by the students.

Although we chose to keep our assessments less formal, providing more formal assessment opportunities for students using this unit would be very simple. For instance, students could be asked to set a goal for the distance their car would move, then document the step-by-step process in reaching this goal. Students could also be required to do a set number of trials and graph the distance moved by their car during each trial. Writing a detailed analysis of the design and movement of their car would provide students with an opportunity to write an analytical paper as part of a cross-curricular lesson. Finally, using photographic evidence of each step in the process, students could use their own pictures in a computer presentation of their project, thus integrating technology into this unit as well.

We discovered so many things in completing this series of lessons with our students. First, and perhaps most important, we found that in allowing the children to design and build their vehicles as they saw fit, we not only captured their attention but also their intrigue. Even the most reluctant students finally jumped on the bandwagon and got hooked. They had the freedom to explore; their only limit was what they set for themselves. It was fascinating to watch the children try SCAMPER and enthusiastically redesign their vehicles before returning to the counter for another trial.

We also noticed that the children took notes, answered questions, made observations, and shared their qualitative data eagerly. Sharing the changes made to their car designs and what resulted from those changes was a constant topic of discussion. For instance, students noted that the candy wheels would not stay on their vehicle, so they shared with their peers how an axle was made using a paper clip. Another student then shared how he had made an axle and added tiny little pieces of tape to serve as a sort of lug nut to prevent the wheels from moving along the axle as the vehicle moved. They didn't feel that they were working because there was an aspect of play in their work. Allowing them to manipulate their materials based on their own observations put them in control. They really were the scientists and engineers. They were instructing themselves. They activated prior knowledge, sought advice and suggestions from classmates, and returned to their work with electricity that you could almost feel.

By the end of the lessons, the students wanted more. Many of the children went home, worked on their designs there, and brought their redesigned air mobiles back in on their own time for a trial run in the library. They did research online, bringing in printouts of engineering sites and air-powered machinery to show us. Many of them shared how they had gotten parents involved to figure out how best to manipulate their vehicles for optimum performance.

What started as a simple way to introduce students to engineering turned into so much more for us and for our students. Their understanding and application of the terms they had learned in their science unit were being transferred to their design challenge. Graphing went from a chore to a competition as they saw which classmates' vehicles went the farthest along the counter. Students even started talking about the SCAMPER tool in their classrooms. With just one simple lesson in a library, learning had more meaning than we had ever imagined!

Connecting to the *Next Generation Science Standards*

The materials, lessons, and activities outlined in this chapter are just one step toward reaching the performance expectations listed below. Additional supporting materials, lessons, and activities will be required.

3-5-ETS1 Engineering Design *www.nextgenscience.org/3-5ets1-engineering-design* **3-PS2-1 Motion and Stability: Forces and Interactions** *www.nextgenscience.org/3-ps2-1-motion-and-stability-forces-and-interactions*	
	Connections to Classroom Activity
Performance Expectations	
3-5-ETS1-1: Define a simple design problem reflecting a need or a want that includes specified criteria for success and constraints on materials, time, or cost	Built a car using a set of materials (four Lifesavers candies, two straws, two paper clips, scissors, tape, and a sheet of paper) and then redesigned the car based on data collected
3-5-ETS1-2: Generate and compare multiple possible solutions to a problem based on how well each is likely to meet the criteria and constraints of the problem	Built a car of their own design that moved with a puff of air; not told how to build the car but designed and redesigned a car until it moved as they desired
3-5-ETS1-3: Plan and carry out fair tests in which variables are controlled and failure points are considered to identify aspects of a model or prototype that can be improved	Used the SCAMPER brainstorming tool to redesign their cars
Science and Engineering Practices	
Developing and Using Models	Constructed models of cars that could be moved by air
Using Mathematics and Computational Thinking	Used observations and discussions to redesign car so it could move farther with air
Disciplinary Core Ideas	
ETS1.C: Optimizing the Design Solution • Different solutions need to be tested in order to determine which of them best solves the problem, given the criteria and the constraints.	Participated in controlled variable tests to collect data about which features of the toy rocket would lead to success or failure
PS2.A: Forces and Motion • Each force acts on one particular object and has both strength and direction. • The patterns of an object's motion in various situations can be observed and measured.	Shared the results of their controlled variable testing with peers to determine which features of the toy rockets would best address the client's needs
Crosscutting Concept	
Patterns	Used patterns of motion to predict future motion

Source: NGSS Lead States 2013.

References

Eberle, B. 1984. *Help! In solving problems creatively at home and school*. Carthage, IL: Good Apple, Inc.

Glen, R. 1997. SCAMPER for student creativity. *Education Digest* 62 (6): 67–68.

NGSS Lead States. 2013. *Next Generation Science Standards: For states, by states*. Washington, DC: National Academies Press. *www.nextgenscience.org/next-generation-science-standards*.

Palincsar, A., and A. Brown. 1984. Reciprocal teaching of comprehension fostering and comprehension monitoring activities. *Cognition and Instruction* 1 (2): 117–175.

Rosenshine, B. 1997. Advances in research on instruction. In *Issues in educating students with disabilities*, ed. J. W. Lloyd, E. J. Kameanui, and D. Chard, 197–221. Mahwah, NJ: Lawrence Erlbaum.

NSTA Connection

Visit *www.nsta.org/SC1509* for a blank version of the SCAMPER brainstorming sheet.

Wacky Weather

An Integrative Science Unit Combines Science Content on Severe Weather With the Engineering Design Process

By Amy Sabarre and Jacqueline Gulino

What do a leaf blower, water hose, fan, and ice cubes have in common? Ask the students who participated in our integrative science, technology, engineering, and mathematics (I-STEM) education unit, Wacky Weather, and they will tell you "fun and severe weather"—words you might not have expected! The purpose of this unit was to interweave science content on severe weather with the engineering design process (EDP). The unit included concepts such as properties of materials, connections to architecture, and the relevance of engineering to our lives.

The EDP is the crux of integrative STEM education. I-STEM education is "the application of technological/engineering design-based pedagogical approaches to intentionally teach content and practices of science and mathematics education concurrently with content and practices of technology/engineering education" (Sanders and Wells 2010). This unit will provide an example of intentional integration of the *Next Generation Science Standards* core ideas and practices (NGSS Lead States 2013).

Designing an Integrative Science Unit

In our school district, the STEM coordinator, elementary teachers, and specialists collaborated to create a model I-STEM education program that gives all students access to the EDP based on science concepts. Goals for this unit included the following:

- Conducting and applying research

- Analyzing charts and graphs of weather data to determine type of severe weather

- Planning, creating, testing, and evaluating their team's design

- Identifying weather instruments and interpreting the measurements

True to the definition of I-STEM education, each unit integrates multiple disciplines so students see the connections between content areas. Units are planned according to the four-part design process: plan, create, check, and share. This simplified engineering and technological design process better fits the elementary realm because it is easy for young children to understand and use. Because all students in our school system participate in the I-STEM units, our units must support the large English language learner population in our district. We are very purposeful in including research-based best practices, such as language objectives, sentence frames, graphic organizers, and strategies to scaffold instruction.

Design Brief

The Wacky Weather I-STEM unit was developed for teachers to use in their classrooms to supplement weather lessons already being taught. Based on science standards, this week-long unit follows the design brief steps of plan, create, check, and share.

Background

In third grade, students study weather and focus on a technical understanding of the tools and methods used to forecast future atmospheric conditions. This unit was designed to follow classroom instruction on weather phenomena, weather measurements, and meteorological tools (performance expectations 3-ESS2-1 and 3-ESS3-1: Earth's Systems and 3-5-ETS1-1 and 3-5-ETS1-2: Engineering Design; NGSS Lead States 2013). Students worked in teams to analyze weather data, predict forthcoming weather phenomena and fortify a previously built structure to withstand an impending "storm."

There are also several *Common Core State Standards (CCSS), Mathematics,* standards integrated into the lessons, including metric measurement, and collecting, organizing, displaying, and interpreting data from a variety of graphs. A large number of *CCSS* language arts standards were also incorporated, such as using effective oral communication strategies, expanding vocabulary when reading, demonstrating comprehension of nonfiction text, and writing for a variety of purposes.

Plan

In the plan phase, students gained the background knowledge needed to be able to complete the design challenge. Observing, exploring, researching, brainstorming and drawing designs all took place in this phase of the process. On day 1, the students were hooked by an opening activity in which they wrote in their design notebook about a time they had personally experienced severe weather (crosscutting concept 3-ESS3-1: Cause and Effect; NGSS Lead States 2013). The timing could not have been better because over the summer a large storm with high winds had knocked down many trees and power lines. Quite a few students wrote about this event and what they observed. This activity naturally differentiated itself because some students wrote a few sentences, others a full paragraph and a few drew pictures and explained to the teacher verbally. Then, students were given time to share with an elbow partner and a few shared their story with the entire class.

Next, the students watched three video clips to review the types of severe weather included in the science standards. The clips focused on thunderstorms, tornadoes, and hurricanes and visually demonstrated the characteristics of each. After students turned and talked with a partner, the teacher recorded characteristics of each type of weather in a whole class chart. The teacher posed questions and tasks such as, "Compare and contrast each type of severe weather." One student responded, "Each storm had wind but the wind made different things happen to the houses."

FIGURE 32.1

Severe Weather Research Sheet

Complete the chart below using information you find on the websites about each type of storm.

Characteristic	Tornado	Hurricane	Thunderstorm
Wind			
Precipitation			
Air pressure			
Other			

After that, students worked in groups of two on iPads or laptops to conduct research using preselected websites that were differentiated by reading level and took notes (see NSTA Connection, p. 282, for a list of websites). Websites were selected based on student accessibility and reading level, engaging content, and the specific content needed for the student research. Several teachers reviewed the websites. Students needing additional support had the research material read aloud to them. All others worked with a partner and recorded information in the data table on characteristics of one type of severe weather (see Figure 32.1 and NSTA Connection, p. 282). At the end of class, students met in groups to share their notes. The students were excited to share with each other about each storm. One animatedly exclaimed, "A tornado has 300 mph winds! That's faster than a racecar!"

During the second day of the unit, students reviewed the four weather instruments by completing a matching activity (see NSTA Connection, p. 282). After demonstrating an understanding of what each instrument was used for, the teacher led an analysis of various charts and graphs with data from weather instruments (see NSTA

Connection, p. 282). Students applied what they had researched and made connections between weather phenomena and the corresponding data weather instruments would measure. One student noted, "Zero centimeters on the rain gauge means there was no precipitation and because the anemometer read 5 m per hour, there was very little wind."

Students were then introduced to the design challenge, constraints, and rubric (Figure 32.2, p. 276, and NSTA Connection, p. 282). Their design constraints were as follows:

Design Challenge: Your team will design and create a structure that will withstand a given form of severe weather (thunderstorm, hurricane, or tornado).

The constraints are as follows:

- Must use recycled materials

- Must be < 40 cm wide

- Must be < 30 cm tall

- Must have a base, four walls, and a roof

- Must include one idea from each member of your team

FIGURE 32.2

Wacky Weather Rubric

Task	Student	Teacher
Group work		
We only talked about our I-STEM project.		
We listened to each other.		
We used kind words.		
Total	/3	/3
Product		
We included recycled materials.		
Our structure was < 40 cm wide.		
Our structure was < 30 cm tall.		
Our structure included a base, four walls, and a roof.		
We included one idea from each person in our design.		
We redesigned our structure to protect it from severe weather.		
Total Points	/6	/6
Presentation		
I made eye contact with the audience.		
I spoke clearly and the audience could hear me.		
I stood up straight.		
Total	/3	/3
Total Points		

Students were allowed time to ask questions about the challenge or constraints. Many students asked questions about the measurement portion to clarify what was meant by wide or tall. Students also asked questions such as, "Do we have to have windows or a door?" The teacher referred the students to the design challenge and asked if this was a constraint. When the students saw that it was not, the teacher then asked, "Do most structures have doors or windows?"

Next, the materials were presented so the students could see what was available and begin brainstorming the design of their structure. Materials included cardboard, Styrofoam, craft sticks, tape, glue guns, and a variety of recycled materials such as water bottles, paper towel rolls and flattened cereal boxes. Students should use caution when cutting materials and using glue guns. Goggles are to be worn during the investigation, cleanup, and hand washing. Close teacher supervision is strongly advised.

The students brainstormed in groups of three and drew a team plan for their structure. They labeled their drawing indicating the materials they planned to use. Each team was reminded to include at least one idea from each team member and to meet with the teacher to get approval for their design plan (disciplinary core ideas ETS1.A: Defining and Delimiting Engineering Problems and ETS1.B: Developing Possible Solutions; NGSS Lead States 2013).

Create

On the third day, students gathered materials and began to build their structure. As students built, the teacher circulated the room asking open-ended questions such as, "Tell me about your design." This question elicits responses that show the teacher areas where guidance or further questioning is needed. For example, "How will your structure support its roof?"

Before the end of class, a scrolling weather alert came across the interactive white board and each group of students was given a column in the severe weather alert chart (Figure 32.3, p. 278, and NSTA Connection, p. 282). Based on this information, the groups had to determine what type of severe weather the information in their column represented. The teacher modeled how to interpret weather data and determine the type of severe weather using an example. Each group met to discuss their data from the severe weather alert and referred to previous pages in their design notebook that held their research from the previous day to determine their storm type. The notebook also contained their meteorological tools matching and their group brainstorming. Next, they brainstormed potential problems their structure might encounter and safety measures they could add to ensure their structure would withstand the severe weather. After discussing, they worked as a team to draw a plan of the modifications they would make to their structure (disciplinary core idea ETS1.B: Developing Possible Solutions; NGSS Lead States 2013).

As the teacher approved each groups' redesign, the students gathered more materials and modified their structures. Students were again asked to return to the criteria and revisit the weather conditions for their type of severe weather. As students built, the teacher moved around the room and asked questions to prompt thought about a specific problem that might occur. For example, the teacher noticed a team with a very light structure and asked them to think about the tornado video they had watched. One student in the team said, "We don't want it to blow away, we have to hold it down." Then, all the team members suggested possible ways to anchor their structure. The students needed all of the class period to modify their structures. Several strategies that our students used during this time were filling water bottles to create weight inside their house; covering their structure with waterproof materials, such as pieces of a plastic trash bag; and placing stakes around the outside to penetrate the ground so the structure would not blow away. Students should use caution when placing stakes because a stake could injure them. Teachers should model pushing a stake through a material by pushing away from their body and carefully supervise students during this process (performance expectation ETS1.B: Engineering and design; NGSS Lead States 2013).

FIGURE 32.3

Severe Weather Alert Chart

Instrument	Yesterday's Forecast	Severe Weather Alert 1	Severe Weather Alert 2	Severe Weather Alert 3
	5 mph	111 mph	164 mph	30 mph
	980 mb	964 mb	900 mb	920 mb
	0 in	9 in	½ in	2 in
	See graph.	See graph.	See graph.	See graph.
Cloud cover	cirrus clouds	stratus clouds	funnel cloud	cumulonimbus
Precipitation	none	heavy rain	none	possible hail

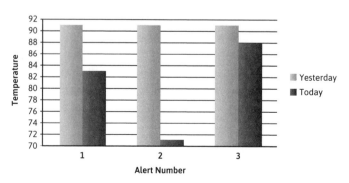

SEVERE WEATHER ALERT

Check

To simulate severe weather, the class used a variety of tools on the fourth day to test the stability of the structures. Leaf blowers were used at close range to simulate the power of the wind in a tornado and at a distance of 30 cm for the hurricane. Teams that determined their severe weather was a tornado also had their homes pelted with ice cubes to simulate hail. Precipitation was further simulated with a water hose for a hurricane and spray bottles filled with water for a thunderstorm. Teachers should closely supervise the testing and should operate all water hoses, fans, and blowers. Electricity and water should be separated. Students were extremely engaged in the testing stage during which cheers and laughter rang out. Picture three girls laying one hand on top of each other and cheering, "3, 2, 1—go, Team Hurricane!" Then, they all throw their hands in the air as if they are at a sporting event. Another group whose structure withstood a "tornado" shouted, "We did it! It survived!"

To facilitate sharing, each group's test was recorded on video. During the testing, teachers continued to ask the groups questions about what they predicted would happen and what they observed. Groups were encouraged to record any problems their structure had and to make the connections between the leaf blower or fan and the force of wind for each type of severe weather.

Share

Allowing students time to communicate their ideas and observations is an essential component of the design process. On day 5, teams met to discuss their results and different modifications they could have made to protect their structure from severe weather. Sentence frames were included in the lessons for students who needed them to organize their thinking. Each team watched their video to observe any additional factors that might have influenced the stability of their structure.

Before presenting their results and reflections to the class, they referred to the rubric. Each team then shared with the class, pointing out unique features of their design, aspects that protected their structure from the severe weather, and thoughts on how they could further improve the structure. The other students enjoyed watching the video clips and actively responded by posing questions and offering suggestions (disciplinary core idea ETS1.C: Optimizing the Design Solution; NGSS Lead States 2013).

Model What You Teach

We have offered an overview of one I-STEM education unit, which can be adapted to different grade levels. I-STEM education requires creative thinking, problem solving, application of knowledge, and collaboration. The process we used with the students is also modeled with teachers in the design and implementation of the units. At the conclusion of the unit, teachers met to reflect on their observations of student performance and engagement and to analyze student design notebooks. The students had an opportunity to self-assess their work using the Wacky Weather rubric. Then, teachers evaluated each design notebook and provided feedback. Based on the evaluation of student work across the grade level, 90% of the students understood the characteristics of severe weather and were able to recognize, name, and explain the types of weather measurements for each instrument.

In addition to reflecting on student performance, teachers also reflected on the lessons and their instructional delivery. When our teachers met to debrief, the overwhelming response centered on the high level of student engagement. They also commented on the noticeable transfer of knowledge that occurred as a result of analyzing the weather data and the range of possible solutions and creativity that the students demonstrated. Most of the suggestions that arose were about tailoring the lessons to meet the needs of individual classrooms. Many teachers shared how they had implemented the share portion of their unit with their students. Although we provided the lessons sequentially, teachers often suggested bringing in parts of the lesson during other content times to allow more time to complete the unit. The integrated curriculum allows for much flexibility during the school day.

All students should have access to high-quality, rigorous, and relevant science instruction. It is this vision that allows our district to continue expansion of STEM programs. The following quote from *Preparing the Next Generation of STEM Innovators: Identifying and Developing Our Nation's Human Capital* encompasses this vision clearly. "The possibility of reaching one's potential should not be met with ambivalence, left to chance, or limited to those with financial means. Rather, the opportunity for excellence is a fundamental American value and should be afforded to all" (NSB 2010). We believe I-STEM education provides this opportunity.

Connecting to the *Next Generation Science Standards*

The materials, lessons, and activities outlined in this chapter are just one step toward reaching the performance expectations listed below. Additional supporting materials, lessons, and activities will be required.

3-ESS2-1 Earth's Systems *www.nextgenscience.org/3ess2-earth-systems* **3-ESS3-1 Earth's Systems** *www.nextgenscience.org/3ess3-earth-human-activity* **3-5 ETS1-1 Engineering Design** *www.nextgenscience.org/3-5ets1-engineering-design* **3-5 ETS1-2 Engineering Design** *www.nextgenscience.org/3-5ets1-engineering-design*	**Connections to Classroom Activity**
Performance Expectations	
3-ESS2-1: Earth's Systems Represent data in tables and graphical displays to describe typical weather conditions expected during a particular season	Represented research of severe weather in a data table including specific characteristics of each storm
3ESS3-1: Earth's Systems Make a claim about the merit of a design solution that reduces the impacts of a weather-related hazard	Justified design solutions and explained the specific modifications that protected the structure from the weather-related hazard
3-5 ETS1-2: Engineering Design Generate and compare multiple possible solutions to a problem based on how well each is likely to meet the criteria and constraints of the problem	Compared each design solution during the share phase of the lesson and reflected on the strengths and weaknesses of each

3-ESS2-1 Earth's Systems	
www.nextgenscience.org/3ess2-earth-systems	
3-ESS3-1 Earth's Systems	
www.nextgenscience.org/3ess3-earth-human-activity	
3-5 ETS1-1 Engineering Design	
www.nextgenscience.org/3-5ets1-engineering-design	
3-5 ETS1-2 Engineering Design	
www.nextgenscience.org/3-5ets1-engineering-design	**Connections to Classroom Activity**
Science and Engineering Practices	
Analyzing and Interpreting Data	Analyzed charts and graphs of weather data to determine type of impending severe weather
Constructing Explanations and Designing Solutions	Explained reasoning for including various modifications to the structure
Disciplinary Core Ideas	
ETS1.B: Developing Possible Solutions • Research on a problem should be carried out before beginning to design a solution. • At whatever stage, communicating with peers about proposed solutions is an important part of the design process, and shared ideas can lead to improved designs.	Researched types of severe weather and recorded data in table Created an individual plan to show modifications that addressed potential problems with the structure, then shared solutions and collaboratively decided on the final design that would withstand the severe weather
ETS1.C: Optimizing the Design Solution • Different solutions need to be tested in order to determine which of them best solves the problem, given the criteria and the constraints.	Tested each structure with a variety of tools to simulate the severe weather, recorded problems with each structure, and then reflected on which structures best solved the problem
ESS2.D: Weather and Climate • Scientists record patterns of the weather across different times and areas so that they can make predictions about what kind of weather might happen next.	Researched weather in the region and recorded data for use in selecting appropriate materials and structures to withstand weather characteristics
ESS3.B: Natural Hazards • A variety of natural hazards result from natural processes. Humans cannot eliminate natural hazards but can take steps to reduce their impacts.	Connected natural hazard impacts and human intervention through video clips, discussion, and the design challenge
Crosscutting Concept	
Cause and Effect	Included the cause and effect of severe weather in a personal narrative

Source: NGSS Lead States 2013.

References

National Science Board (NSB). 2010. *Preparing the next generation of STEM innovators: Identifying and developing our nation's human capital*. Arlington, VA: National Science Foundation.

NGSS Lead States. 2013. *Next Generation Science Standards: For states, by states*. Washington, DC: National Academies Press. *www.nextgenscience.org/next-generation-science-standards*.

Sanders, M., and J. Wells. 2010. Integrative STEM education. Virginia Tech School of Education. *http://web.archive.org/web/20100924150636*.

Internet Resource

Downloadable Smart Board Activities
https://docs.google.com/a/harrisonburg.k12.va.us/file/d/0B_dUrNTXooKGVjYtX2xQUXZyUkE/edit

---

NSTA Connection

Visit **www.nsta.org/SC1310** for an Internet safety note, a list of preselected severe weather websites, a severe weather research sheet, weather instruments matching activity sheet, sample chart and graph with data from weather instruments for student analysis, rubric, and severe weather alert chart.

Straw Rockets Are Out of This World

STEM Activities for Upper-Elementary Students

By Joan Gillman

What child does not gaze at the night sky, held spellbound by the awesome sights above? How many times have your students dreamed of going on a rocket ship and visiting the planets in our solar system? To capture this excitement and engage students' interest, I have designed lessons that give students the opportunity to experience the joys and challenges of developing straw rockets and then observing which design can travel the longest distance. The lessons are appropriate for students in grades 4 and 5. I have used the rocket-building unit very successfully with my fifth-grade students. It is such a joy watching my students working diligently on the investigation while at the same time enjoying the challenges the unit presents.

Our straw rocket investigations relate to disciplinary core idea PS2.A, as explained in the grade 5 endpoint found in *A Framework for K–12 Science Education,* "Each force acts on one particular object and has both strength and a direction. An object at rest typically has multiple forces acting on it, but they add to give zero net force on the object. Forces that do not sum to zero can

cause changes in the object's speed or direction of motion" (NRC 2012, p. 115). Although our experiences are with fourth and fifth graders, the *Next Generation Science Standards* identify a third-grade standard, 3-PS2 Motion and Stability: Forces and Interactions, which addresses performance expectations 3-PS2-1: "Plan and conduct an investigation to provide evidence of the effects of balanced and unbalanced forces on the motion of an object" and 3-PS2-2 "Make observations and/or measurements of an object's motion to provide evidence that a pattern can be used to predict future motion" (NGSS Lead States 2013).

In addition, a related performance expectation is engineering design 3-5-ETS1-3: "Plan and carry out fair tests in which variables are controlled and failure points are considered to identify aspects of a model or prototype that can be improved" (NGSS Lead States 2013). The experiences also support the understanding of the crosscutting concepts of patterns and cause and effect.

The straw rocket unit fits very well with these standards. In this unit, the students will be working with both controlled and manipulated variables. They will be testing and evaluating each variable to determine the ideal flying design for their straw rocket. Throughout this unit, the students will be actively engaged in the learning process. They will be looking at the flight data they collect. They will be analyzing the results of their straw rocket flights, and they will be redesigning their rockets to improve their original design.

A Brief History of Rockets

To begin this unit, the students look at the history of rockets. From Chinese fireworks to the development of modern rockets, the history of rocket construction is a fascinating story. In the 1860s, the science fiction writer Jules Verne envisioned a spacecraft that could be shot out of a cannon and sent to the Moon. If you go further back in time, the first rockets were actually developed by the Chinese. There is a famous Chinese legend that dates from approximately 1500, which is about an official named Wan Hu. He attempted to fly to the Moon by tying a number of rockets to his chair and lighting them. The rockets made a huge roar when they exploded. Once the smoke had cleared, there was no trace of either Wan Hu or his chair. If we jump ahead to the 1900s, we can see the initial development of modern rockets. Much of this development can be attributed to scientists such as Russian physicist Konstantin Tsiolkovsky, American physicist Robert Goddard, and German rocket designer Wernher von Braun. There are some excellent trade books that can be used to help students advance their knowledge of the history of space flight. Two of them—*The History of Rockets* (1999) and *Rockets* (2008) are written by Ron Miller.

How Rockets Work

Before the students begin designing and building straw rockets, they need to comprehend what a rocket is and how it works. I introduce the lesson with a brainstorming session. The first question asked is, "What is a rocket?" Student answers have included, "It's a device that takes people to outer space" or "It's a large, noisy machine that lifts off the ground with lots of gas and smoke." The next question is usually more challenging: "What do you think enables a rocket to fly?" Responses have included, "It uses lots of fuel" or "A large explosion lifts the rocket off the ground." Once we've concluded this session, we're ready to begin discussing the science behind rockets.

This unit exposes children to Newton's third law of motion. Newton's third law describes the relationship between the forces that two objects exert on each other. Forces always come in equal but opposite pairs. An easy way to demonstrate this would be with a balloon. If you fill a balloon with air and then release it, air escapes and the balloon is pushed in the other direction. The force exerted by the balloon on air moving out of the balloon is an action force. An equal force, exerted by the air moving out of the balloon, pushes the balloon forward. This is the reaction force. This activity really excites the children. What child has not tried this same demonstration at home? You can also extend this lesson by making a balloon rocket using a balloon, kite string, a plastic straw, a clear plastic bag, and tape. If there are students with latex allergies, you may want to consider purchasing balloons made from vinyl, foil, or plastic (see NSTA Connection, p. 290, for a description of how to conduct this activity and places to purchase various kinds of balloons).

The main feature of a rocket is its ability to expel gas in one direction. When gases shoot out of the back of the rocket, the rocket is pushed in the opposite direction. For every force or action in which the rocket pushes back on exhausted fuel gases, the exhausted fuel gases exert an equal and opposite force or action on the rocket. The reaction force that propels a rocket forward is called thrust. The amount of thrust depends on a few factors, including the mass and speed of the gases propelled out of the rocket. The greater the thrust, the greater a rocket's velocity. For the rocket to lift off the ground, it must experience more upward thrust than the downward force of gravity. A good trade book I use in the classroom to illustrate these concepts is *Simon Bloom: The Gravity Keeper* (Reisman 2009). The text discusses gravity, inertia, friction, and Newton's laws of motion. The energy source for the rocket is the straw rocket launcher (see Resources). The straw rocket launcher can be expensive. An alternative would be to construct your own. See "The Ten Dollar Rocket Launcher" online for step-by-step instructions for making your own device (Jogerst 2008).

To launch the rockets, the straw rocket launcher uses the pneumatic force created by releasing a weighted drop rod in the cylinder. The force of the launch can be controlled by varying the release height of the rod. There are some recommended trade books that can expand the students' comprehension of how rockets work. These include *How Does a Rocket Work?* by Sarah Eason (2012), *Rockets and Other Spacecraft* by John Fardon (2000), and *Master Engineer: Rockets* by Paul Beck (2010).

Let's Begin the Straw Rocket Investigation

In the NSTA Connection (p. 290), you can find the directions for the straw rocket activities as well as the recording and data sheets. Why should you choose to use straw rockets in the classroom? It's an excellent way to include STEM skills in your curriculum. You'll be combining science, model building, data collection, design testing, and some math skills in this project.

When I begin the building aspects of this curriculum, I show the video *Dr. Zoon Presents Straw Rockets* made by Pitsco Education. It provides an excellent introduction to building and testing straw rockets. The original "Dr. Zoon" video has now been replaced with one called "Straw Rocket Video" (see Internet Resource, p. 290).

After showing the video, I commence the next lesson with the question, "How can you design a straw rocket to reach the farthest distance in the path that you expect the rocket to take?" Before discussing the rocket's specifications, I spend a few minutes seeing what ideas the students might have for their rockets. Some children have imagined having triangular or square-shaped rockets. Others have described rockets that are similar to the ones at NASA. Once the discussion concludes, we are ready to begin the formal planning of the rockets. Because we'll be working with specific variables, it's important to review the specifications for the rocket (see NSTA Connection, p. 290, for the direction sheets). The straw length may be 10–20 cm. There can be 2–5 fins, and the clay nose cone must be 2 cm or less in diameter. Finally, the students select and record three launch angles to test by adjusting the angle mechanism on the launcher. The same rocket will be used for the three different launch angles. Once we have reviewed the specifications, students record their hypotheses and begin planning their rockets. This first part takes the form of a sketch of their rockets. Next, the students begin building their rockets. At this stage, measurement skills become important. Students use rulers to check the length of the straw and the diameter of the nose cone and use a triple beam balance to determine the mass of the rocket. If students are more familiar with a double pan balance, this can be substituted. Attaching fins to the straw body can be challenging. I recommend pairing the students with a partner that has facility with this skill. Because I try to instill in my students the philosophy that we are a community of learners, I encourage the children to help one another.

Straw Rocket Testing

Once the rockets are designed, built, and measured, we're ready to begin testing the rockets. This can be accomplished outside in an open space. It is possible to do the activity inside, but you will need sufficient space. My student's record for the longest flight was 39 m and 90 cm! Because the students have chosen three different launch angles, they'll have the opportunity to record the distance the straw rocket travels for each of the chosen launch angles.

When testing the rockets, choose a day when the weather is fair with little wind. The rocket's mass is very small, so any amount of wind or wet conditions will affect its flight.

Before testing the rockets, I recommend securing about 30 m of measuring tape to the ground. This will enable the students to determine how far their rockets have flown.

Don't forget to have the students wear their safety goggles and remain clear of the launch area. The flight path of the rocket can vary greatly depending on its construction, so there is the possibility that it could fly off to the side and hit one of the students. I highly recommend making sure the flight path is clear before the students set off their rockets.

Next, have the students bring the recording data sheet, a clipboard, and a pencil so that they can record their rockets' flight distances (see NSTA Connection, p. 290, for the data sheets). As the rockets are set off, it's easy for the children to get caught up in the excitement of the activity and forget to record flight distances. A few reminders should help keep them focused on their tasks. Expect to hear shouts of joy as the rockets soar through the air. There will also be some disappointment because not all rockets will fly well. You can reassure the students that they will have another opportunity to improve their rocket designs.

The next step is to analyze the results of the rockets' flights (see NSTA Connection, p. 290, for the analysis sheets).

At this point, the only variable tested was the launch angle. The students haven't tried changing the rocket designs to see how that would affect the flight distance. It's possible to set up a more systematic approach using controlled and manipulated variables. In the initial investigation, the controlled variables were the length of the straw body, the number and shape of the fins, and the size of the nose cone, whereas the manipulated variable was the launch angle.

Straw Rocket Building Continued

In the following investigations, the students can manipulate one of the previously controlled variables while keeping the other ones controlled. For example, the first manipulated variable could be the length of the straw body. Students would design, build, test, and record data for the flights of a 10 cm, 15 cm, and 20 cm straw rocket. At the same time, the number and design of the fins would have to remain the same, the size of the nose cone would need to be uniform, and the launch angle would have to stay at 45°. By keeping the other variables controlled, students would be able to interpret how each manipulated variable affects the flight of the rocket.

The next manipulated variable could be the number of fins, and the final manipulated variable could be the size of the nose cone (see NSTA Connection, p. 290, for the manipulated variable direction sheets).

At the conclusion of the unit, the students should be able to come to a consensus on the type of straw rocket that goes the farthest.

Straw Rocket Building Directions

Here are some easy-to-follow directions for building the straw rockets:

1. Take one of the straws and cut it so that the length is between 10 cm and 20 cm.

2. On an index card, draw one of the fins you will be attaching to the rocket. Cut out the fin and use that as a template to make the additional 1–4 fins.

3. Attach the fins to the straw rocket using transparent tape.

4. Construct a clay nose cone 2 cm or less in diameter.

5. Place the clay nose cone ½ cm into the end of the straw.

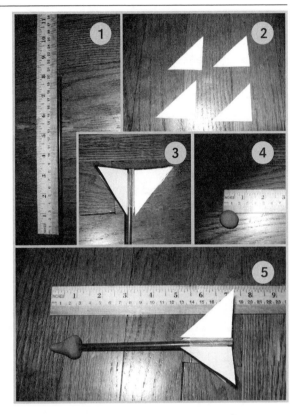

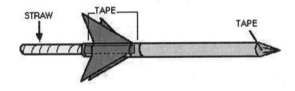

Some Misconceptions

When students are introduced to the rocket building and launching topics, they may have a few misconceptions or misunderstandings. Initially, the children become fixated on the aesthetic appearance of the rockets rather than thinking about what qualities might enable the rockets to fly the longest distance. This misunderstanding usually becomes less of an issue once the students analyze their data after their rockets' first flights. A second area of difficulty involves accurately measuring and weighing their rockets. Some students use the wrong side of the ruler and report measurements in customary units rather than metric ones. Finally, using a triple beam balance or double pan balance to find the mass of the rocket can be challenging.

Young children can also find it challenging to analyze data and write conclusions based on their findings. This is a vital skill—one that they will be using throughout their lives. In the fall, I frequently model how to write a lab report. Once the rocket building unit commences, students have begun to develop some facility with this skill. This becomes especially important because they will need to examine the results of their rockets' flights and then manipulate the variables so that they can enhance their rockets' ability to fly.

Lesson Evaluation

In the straw rocket unit, I used a variety of methods to assess student learning. During one class, students were asked to plot the results of the previous day's rocket flight. First, I looked to see that the students had accurately noted the distances using the metric system. Next, I observed the students to see if they were struggling with the printed directions. I also checked whether they correctly measured and built their rockets using metric measurements and if they had developed more facility with use of the triple beam balance or double pan balance. Next, I evaluated how well the students could articulate what they were doing. Finally, I wanted to see if the students could build a more successful second rocket based on the results of their first model. I have developed a rubric to use when evaluating the students' performance in this investigation (see NSTA Connection, p. 290).

The straw rocket unit can be an exciting and educational experience. When asked at the end of the year what their favorite activity was, students usually said the straw rocket unit. I personally enjoy teaching this unit because it's a great way to integrate STEM skills into the curriculum while getting students excited about the learning process.

Connecting to the *Next Generation Science Standards*

The materials, lessons, and activities outlined in this chapter are just one step toward reaching the performance expectations listed below. Additional supporting materials, lessons, and activities will be required.

3-PS2 Motion and Stability: Forces and Interactions www.nextgenscience.org/3ps2-motion-stability-forces-interactions **3-5-ETS1 Engineering Design** www.nextgenscience.org/3-5ets-engineering-design	**Connections to Classroom Activity**
Performance Expectations	
3-PS2-1: Plan and conduct an investigation to provide evidence of the effects of balanced and unbalanced forces on the motion of an object	Designed and built straw rockets to determine how the structure of the rocket affects its flight
3-PS2-2: Make observations and/or measurements of an object's motion to provide evidence that a pattern can be used to predict future motion	Observed and recorded the flight distance of the straw rockets to provide data to predict future flights of the rockets
3-5-ETS1-1: Define a simple design problem reflecting a need or want that includes specified criteria for success and constraints on materials, time, or cost	Given specific materials (straw, modeling clay, sheet of paper, index cards, scissors, and clear tape) to design and build a straw rocket to travel the furthest distance when launched

3-PS2 Motion and Stability: Forces and Interactions www.nextgenscience.org/3ps2-motion-stability-forces-interactions **3-5-ETS1 Engineering Design** www.nextgenscience.org/3-5ets-engineering-design	**Connections to Classroom Activity**
Performance Expectations	
3-5-ETS 1-3: Plan and carry out fair tests in which variables are controlled and failure points are considered to identify aspects of a model or prototype that can be improved	Tested straw rockets to see which models performed the best, identified what aspects needed to be improved, and refined models to meet those specifications
Science and Engineering Practices	
Planning and Carrying Out Investigations	Designed several options for their rocket based on the parameters provided and chose a model to construct and test Used quantitative results that were collected and analyzed
Constructing Explanations and Designing Solutions	Determined which model provides the straw rocket with the best chance for a longer flight distance; redesigned, rebuilt, and tested straw rockets to prove which criteria provides the best flight results
Disciplinary Core Ideas	
PS2.A: Forces and Motion • Each force acts on one particular object and has both a strength and a direction. An object at rest typically has multiple forces acting on it, but they add to give zero net force on the object. Forces that do not sum to zero can cause changes in the object's speed or direction of motion. • The patterns of an object's motion in various situations can be observed and measured; when the past motion exhibits a regular pattern, future motion can be predicted from it.	Participated in several forces-related investigations, analyzed results, and applied information about forces to the straw rockets Collected data from all investigations and analyzed results for patterns that may affect the motion of straw rockets
ETS 1.B: Developing Possible Solutions • Research on a problem should be carried out before beginning to design a solution. Testing a solution involves investigating how well it performs under a range of likely conditions. • Tests are often designed to identify failure points or difficulties, which suggest the elements of a design that need to be improved.	Studied history of rocketry, how rockets are propelled, and Newton's third law before designing, building and launching a straw rocket to go the furthest distance. Repeated multiple tests on a variety of variables and recorded results for analysis

3-PS2 Motion and Stability: Forces and Interactions www.nextgenscience.org/3ps2-motion-stability-forces-interactions **3-5-ETS1 Engineering Design** www.nextgenscience.org/3-5ets-engineering-design	**Connections to Classroom Activity**
Disciplinary Core Ideas	
ETS 1.C: Optimizing the Design Solution • Different solutions need to be tested in order to determine which of them best solves the problem, given the criteria and the constraints.	Analyzed the results of the flight and, as a class, determined what factors were needed to improve the distance the rocket traveled
Crosscutting Concepts	
Patterns	Used the patterns of motion witnessed during rocket flights to predict future motion
Cause and Effect	Explained how changing the size and shape of the rocket fins, the size and weight of the nose cone, and the length of straw affects the fight of the rocket

Source: NGSS Lead States 2013.

References

Beck, P. 2010. *Master engineer: Rockets.* Charlotte, NC: Silver Dolphin Books.

Eason, S. 2012. *How does a rocket work?* New York: Gareth Stevens Publishing.

Fardon, J. 2000. *Rockets and other spacecraft.* Kent, UK: Copper Beech Publishing.

Jogerst, J. 2008. The ten dollar rocket launcher: Simple science and cheap thrills. Air Force Association, Hurlburt Field Chapter 398. *http://papermodelingman.com/straw09/Straw%20Rocket%20Lesson%20Material%20with%20BIG%20IDEA%20STD.pdf.*

Miller, R. 1999. *The history of rockets.* New York: Franklin Watts.

Miller, R. 2008. *Rockets.* Minneapolis: Twenty-First Century Books.

National Research Council (NRC). 2012. *A framework for K–12 science education: Practices, crosscutting concepts, and core ideas.* Washington, DC: National Academies Press.

NGSS Lead States. 2013. *Next Generation Science Standards: For states, by states.* Washington, DC: National Academies Press. *www.nextgenscience.org/next-generation-science-standards.*

Reisman, M. 2009. *Simon Bloom: The gravity keeper.* New York: Puffin Books.

Internet Resource

Pitsco Education: Straw Rocket Video *www.pitsco.com/store/detail.aspx?ID=6372&bhcp=1*

Resources

Asyby, R., and R. Hunt. 2004. *Rocket man: The Mercury adventure of John Glenn.* Atlanta: Peachtree Publishers.

Prentice Hall. 2007. *Science explorer: Astronomy.* Upper Saddle River, NJ: Prentice Hall.

Otfinoski, S. 2006. *Rockets.* New York: Benchmark Books.

Sobey, E. 2006. *Rocket-powered science: Invent to learn! Create, build, and test rocket designs.* Culver City, CA: Good Year Books.

NSTA Connection

Visit **www.nsta.org/sc1310** for the lesson materials.

34

Nature as Inspiration

Learning About Plant Structures Helps Students Design Water Collection Devices

By Kristina Tank, Tamara Moore, and Meg Strnat

A few weeks earlier, if you had asked the students in Meg Strnat's fourth-grade class to tell you about the issue of water shortage, you would have received blank looks. However, if you had visited their classroom on this particular day, you would have seen and heard a different story. Students huddled around their water collection designs preparing for their final presentations. A group in the front was practicing, "Our final design was inspired by a tree with droopy leaves where the water slips off and drips to the ground." Another group could be heard saying they improved their design by adding a funnel that was inspired by the cup-like shape of flowers. These students were well-versed in explaining how their designs solved the engineering problem and had been inspired by nature, but this hadn't always been the case.

Meg's class was engaging in an engineering design unit in which the students had been asked by an Engineers Without Borders chapter (their client) to help them design water storage and collection devices for families on Popa Island in Panama. When they started this unit, Meg noticed her students had been really

engaged when it came to the water collection part of the challenge, which was to use plants as inspiration for gathering water. However, they were having a hard time grasping the context of this engineering problem—that there are places without running water. After realizing their difficulty in relating to this engineering challenge, Meg decided she needed to help her students better understand the problem of water shortage and why it was important around the world before continuing with the engineering challenge. Meg's recognition of the need for students to clearly understand the engineering problem is an important idea that is captured in the engineering standards in the *Next Generation Science Standards* (NGSS Lead States 2013).

This chapter describes the final lesson in a seven-day STEM and literacy unit that is part of the Picture STEM curriculum (see Internet Resource, p. 297) and uses engineering to integrate science and mathematics learning in a meaningful way (Tank and Moore 2013). For this engineering challenge, Meg's students used nature as a source of inspiration for designs to collect and store water for the people of Popa Island by highlighting that plants have physical structures that help them better collect and store water.

Background

Before diving into the science that students would need for the engineering design challenge, Meg introduced the concept of biomimicry and how engineers can use nature as inspiration for solving problems. She presented students with pieces of Velcro and asked them, "Where do you think the idea for Velcro came from?" to think about the inspiration for inventions. Students were surprised to learn that Velcro came from those annoying burs found in their backyard. Students continued to learn about biomimicry as they explored how bird's wings inspired airplane wings and how the sticky hairs on gecko's feet could help with the development of stronger adhesives.

Once this idea of using nature as inspiration for engineering design had been developed, students explored the science content of how plant adaptations help organisms survive and learned that these adaptations can inspire engineering solutions. The adaptation lessons focused on physical structures that provide advantages. Meg had her students research different biomes from around the world with a focus on the plants that live in these particular biomes and what adaptations helped the plants survive in that biome. To help students make connections between the science content and the engineering design challenge, Meg asked students to create biome posters showing the location and features of the biome as well as examples of plants that lived there and what adaptations helped the plants survive in that biome (Figure 34.1). During the presentation of their posters, students shared how these adaptations could be used in the design of their water storage and collection devices.

Evan, whose group has studied the desert biome, shared that "plants have really long

FIGURE 34.1

Biome Poster

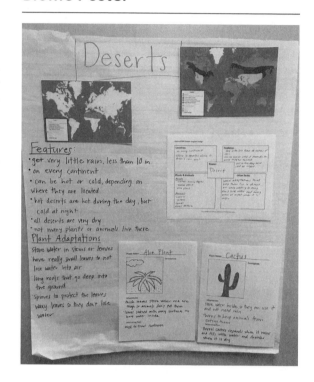

roots to help suck up the water, so we could use sponges in our designs to suck up the rain water." Maricela said, "Many plants have leaves that are called drip tips that look like funnels, so we could put a funnel on our storage tank to get more water."

Understand, Plan, and Create

Meg started the next lesson with a few pages from the picture book *One Well, the Story of Water on Earth* (Strauss 2007). She chose this book because it would help her students grasp the idea that there is a finite amount of water on Earth and that access to clean water is not the same for everyone. She also wanted to help them understand how much water is needed for simple tasks such as washing your hands, so she had her students collect water in buckets that were placed in the sink. Instead of letting the water run down the drain, the buckets captured the water from each group as they washed their hands. Then, each group measured the amount of water that collected in their bucket, and they added that information to a class chart, as well as the total amount used, before calculating the average amount of water used each time someone washed his or her hands. This visual stayed on the wall for the remainder of the unit and was used as a reference when talking about the problem of water shortage and how much water is needed for various tasks.

Equipped with the necessary background knowledge, Meg knew her students were almost ready to start on the engineering design challenge. To help prepare for the challenge, Meg placed students into their predetermined science groups, which consisted of four students with varying strengths and abilities. She reminded them about the expectations for working together as engineers, such as encouraging each other if they notice that someone's ideas are not being heard. However, she first wanted to make sure her students had a good understanding of the problem they were solving. As a formative assessment of the students' understanding of the engineering challenge, she asked the students to individually describe the problem they were trying to solve. Then, she had them share it in their groups and concluded with a whole-class discussion. The individual brainstorming allowed students the chance to think of their own ideas, which helped to ensure that everyone could bring an idea to discuss with their groups. When the students and groups were able to define the engineering problem and explain how they were trying to help solve the problem, Meg knew they were ready to move on. Figure 34.2 shows the problem statement of one of the small groups.

Another important part of understanding the engineering problem is the ability to identify

FIGURE 34.2

Problem Statement

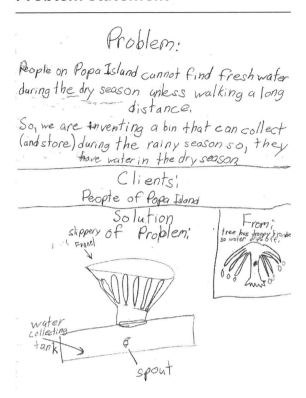

and consider the constraints and trade-offs when brainstorming possible solutions and making individual and group design plans (NGSS Lead States 2013). To help develop this type of thinking, Meg led a class discussion about how engineering problems rarely have a single solution. For this challenge, the storage tank needed to be large enough to store the water they would need during the dry season. However, if the tank was too big then it would cost more and take up more space in the family's backyard. To help make this idea of trade-offs more concrete for her students, Meg presented the testing criteria that they would be using and highlighted that designs could receive points and be successful in many different ways. Each design received points for the volume and how well it matched with the amount of water needed, the cost of their materials, the design's performance on the two water collection tests, and if their design was inspired by nature. To further develop this idea of trade-offs, Meg had students identify good and bad reasons for using different storage container sizes, such as cost. Then, groups had to decide which container they wanted to use and explain to the class why their group decided on that container size. Once the groups had decided on their storage container size, Meg reminded her students that the other constraint that they needed to address came from their client; designs needed to be inspired by plant adaptations. This use of the adaptations allowed the designs to replicate physical structures that plants use to improve the efficiency of their water collection and storage designs.

You could see and feel the excitement in the classroom through the buzz of voices as students quickly moved from identifying the problem to brainstorming possible designs. Meg asked students to identify which size collection device their group chose and which adaptations they would consider in their initial designs. Doing so helped students to be more intentional in the application of the science and mathematics content that they learned in previous lessons, instead of jumping right into the first idea that came to mind. As Meg walked around the room, she heard students talking about how they would use the science and mathematics learning from the past few lessons in the brainstorming of their designs. One group was talking about the size and volume of the storage containers they chose as they calculated their volume score and their cost of materials (which is related to surface area). With the science learning, one student remarked, "I used the Bromeliad because its shape helps collect and make water go into the collection tank." This type of evidence-based reasoning was heard throughout the classroom as groups shared their initial designs and worked to try and come to a

single group design. While this step of combining ideas can be difficult for students, it is important to allow them to share their own ideas as well as hear about different design possibilities to reinforce the idea that there are multiple ways to address engineering problems.

Creating and Testing

Once the groups agreed on a final design, they were ready to start the creating and testing phase of the design challenge. Students will be using scissors to cut various materials during this step. One of the struggles Meg noticed during the "create" phase was that even as students were nearing completion of their designs, they wanted to make changes from their initial design. The group with the Bromeliad-inspired design could be overheard talking about how if they changed the "leaves" of their design to one continuous piece then it would catch more rain. However, to help students complete controlled tests of their agreed-upon group design, Meg encouraged groups to follow their initial plans and reminded them they would have a chance to change things during their redesign. This helped to direct students' reasoning for potential improvement during their redesign to be based on testing and data. After completing their initial designs, Meg had groups conduct two water collections tests that helped groups to determine the success of their designs based on the points they received from how well they met the criteria and constraints.

Sharing and Redesign

To differentiate between the groups who were still finishing their designs and those who had completed the testing, Meg allowed the groups that had tested to start on the redesign phase. In recognizing the importance of redesign for student learning, Meg helped groups identify how

FIGURE 34.3

Engineering Notebook Entry

Nature Inspired Design – Evaluate

3. What worked well with your design?
The hole in the middle helped because it brought in more water.

4. What didn't work well with your design?
The aluminum foil because it leaked and seeped through.

5. What did you see in another group's design that you liked?
We saw a funnel that brought the water down into a straw.

6. What are some ideas for how you could improve your design for next time?
Use popsicle sticks instead of aluminum foil for more sturdiness to hold water.

7. For our redesign we decided to change:
We decided to change the base because tin foil was weak so we changed it to foam.

8. We think this will help because:
foam is sturdier than tinfoil. (Not light)

and why their proposed changes would improve their designs by completing an evaluation sheet in their engineering notebooks. A student example of one engineering notebook page is shown in Figure 34.3. The engineering notebook was integrated throughout the unit as a formative assessment and allowed Meg to assess student's understanding as they progressed through the unit. After each of the groups had a chance to redesign, final designs were shared by creating videos that explained the designs, test results, and the way the design had been inspired by nature. Figure 34.4 shows one group's initial design and students working on their redesign.

FIGURE 34.4

A Student's Initial Design (left) and Redesign (right)

Conclusion

The engineering design component of this STEM unit made it possible to tie students' mathematics and science learning together in a real-world context that highlighted problem solving as well as generating and testing multiple solutions. Meg commented throughout the unit that she was impressed by the excitement and engagement of her students as they worked toward their final designs. She was also encouraged by the fact that her students had made meaningful design considerations using the mathematics and science they had learned as part of this unit, which she could see as they explained their designs in their final product presentations. When thinking about how to include engineering in elementary classrooms, picture books can be used to help students better understand the context of engineering design problems.

Connecting to the *Next Generation Science Standards*

The materials, lessons, and activities outlined in this chapter are just one step toward reaching the performance expectations listed below. Additional supporting materials, lessons, and activities will be required.

3–5 ETS1 Engineering Design *www.nextgenscience.org/3-5ets1-engineering-design*	Connections to Classroom Activity
Performance Expectations	
3-5-ETS1-1: Define a simple design problem reflecting a need or a want that includes specified criteria for success and constraints on materials, time, or cost	Explained the engineering problem, including the criteria and constraints
3-5-ETS1-2: Generate and compare multiple possible solutions to a problem based on how well each is likely to meet the criteria and constraints of the problem	Individually created multiple solutions to the design challenge before working together to evaluate how their ideas meet the specific constraints and criteria to create a single group design.
Science and Engineering Practices	
Asking Questions and Defining Problems	Explained the engineering problem
Planning and Carrying Out Investigations	Created and tested group design (and redesigns) to determine how well the design fits with the initial constraints and criteria
Constructing Explanations and Designing Solutions	Constructed explanations regarding the success of the final design and the steps taken to get to these designs
Disciplinary Core Ideas	
TS1.A: Defining and Delimiting Engineering Problems • Possible solutions to a problem are limited by available materials and resources (constraints). The success of a designed solution is determined by considering the desired features of a solution (criteria). Different proposals for solutions can be compared on the basis of how well each one meets the specified criteria for success or how well each takes the constraints into account.	Learned about the engineering design problem, brainstormed solutions, and worked to come to a single solution that best fits the specified criteria
K-ESS.C: Human Impacts on Earth Systems • Things that people do to live comfortably can affect the world around them.	Gathered evidence of water use in daily life and the way individual usage impacts others' usage
Crosscutting Concept	
Systems and System Models	Created a model of a system used to collect and store water that can be tested in the classroom

Source: NGSS Lead States 2013.

Acknowledgment

The material reported in this chapter was supported by the National Science Foundation under Grants No. EEC–1442416. Any opinions, findings, and conclusions or recommendations expressed in this material are those of the authors and do not reflect the views of the National Science Foundation.

References

NGSS Lead States. 2013. *Next Generation Science Standards: For states, by states*. Washington, DC: National Academies Press. *www.nextgenscience.org/next-generation-science-standards*.

Strauss, R. 2007. *One well: The story of water on Earth*. Toronto: Kids Can Press.

Tank, K. M., and T. J. Moore. 2013. *Nature-inspired design: A grades 3–5 PictureSTEM curriculum unit*. Developed from National Science Foundation grant EEC-1442416. Washington, DC: American Society for Engineering Education.

Internet Resource

PictureSTEM
http://pictureSTEM.org

The Tightrope Challenge

When Confronted With a Robotics Engineering Task, Fourth-Grade Students Develop Growth Mindsets

By Bill Burton

In truly authentic problem-solving challenges, teachers may not know what will happen or how a final project will look. When teachers give up some control of an engineering project and hand it over to students, some amazing things can happen.

To prepare our students to become our next innovators, teachers need to provide real-world challenges that allow children to exercise their innovation muscles. The real-world innovation process doesn't happen on a worksheet, and it doesn't come with a detailed set of directions. Innovation starts with a problem and innovators work to solve a problem by planning, creating, and testing. Along the way, there may be successes and setbacks, joy and frustration, teamwork and friction.

Often, the setbacks and frustrations in a project lead to the greatest successes and the most meaningful learning. Thomas Edison tested more than 1,600 materials for the filament of the lightbulb before he found one that worked (Bedi 2004). As innovators, our students need to have Edison's resiliency.

In her book *Mindset*, Carol Dweck (2008) coined the terms *growth mindset* and *fixed mindset*. Dweck states that people with fixed mindsets believe that certain personal qualities are static. They cannot grow beyond who they are. When met with a challenge that doesn't offer initial success, they may be quicker to make excuses, assign blame, or simply give up. However, Dweck describes people with growth mindsets as having resilient characteristics. They relish a challenge. When faced with a setback, they think of ways to improve their performance. They have a resiliency that innovators need.

When my students are given an engineering challenge, the key word is *challenge*. Challenges that aren't challenging don't help students learn new skills or try new things. It can quickly become clear which type of mindset a student has during a particularly difficult challenge. But what's great about Dweck's mindsets is that they can change. As educators of innovators, it is our role to nurture the growth mindset.

Background

Lego Mindstorms robotics sets offer endless opportunities for students to complete engineering and programming challenges. Before entering fourth grade, students at our school have had significant experience with both Mindstorms and WeDo Lego systems and programming. In addition, classes have spent time discussing the steps of the engineering or innovation process. While this chapter demonstrates the engineering process through the lens of Lego products, there are certainly other activities that require less investment in materials (see "Alternatives to Legos," p. 303).

The Engineering Process

There are many resources that discuss the engineering process. The steps of the engineering process can take on many forms. In several cases, the engineering process itself depends on the problem being solved. Although the process isn't bound by any rules, it generally includes the following six steps (in some varied order or detail):

- Problem Statement

- Research

- Design

- Create

- Test

- Improve

These steps are often cyclical in nature. A design may be built and tested. Then, improvements need to be made that require more building and testing. Several websites offer great kid-friendly graphics describing the engineering process (see Internet Resources, p. 306).

Setting up the Challenge

In fourth grade, students are given several open-ended robotics challenges. For the purpose of this chapter, the tightrope challenge will be discussed. In the tightrope challenge, a length of rope was suspended at student height and secured at each end of the room. A teacher-designed Lego device secured at one end of the rope held a plastic ball (see Figure 35.1 and NSTA Connection, p. 306, for directions). The ball holder can also be made from materials such as cardboard and tape, as long as it serves the purpose of holding a ball without fully enclosing it.

FIGURE 35.1

Photograph of Lego Device on Tightrope

Before dividing the students into smaller groups, the class discussed a range of possible achievement levels for the challenge. At minimum, each group was required to design and build a vehicle that could suspend from the rope and move in some way. The most ambitious achievement level challenged students to design, build, and program a robot to traverse the rope, stop at a designated point, retrieve a plastic ball, stop at another designated point, and then drop the ball. Students were also given the option to create a project aimed somewhere between the two extremes of the achievement spectrum. For example, some groups might choose to spend more time designing a project to retrieve the ball more efficiently. Or, others decided to try unique designs to traverse the rope in a fun way. The initial task was set with the potential for students to make it their own.

Planning for the Challenge

Although students often want to dive right into building, project planning is helpful. For the tightrope challenge, fourth-grade students did basic project planning. For the first step, each student defined the problem that had been presented. Students wrote a problem statement on an engineering planning sheet (see Figure 35.2 and NSTA Connection, p. 306). This was a great way to assess if they understood what was expected and what their final project should accomplish. This aligns with the *Next Generation Science Standards* engineering design performance expectation 3-5-ETS1-1: "Define a simple design problem reflecting a need or a want that includes specified criteria for success and constraints on materials, time, or cost," as well as disciplinary core idea ETS1.A: Defining and Delimiting Engineering Problems (NGSS Lead States 2013). Students were then expected to draw a sketch of what their project might look like (Figure 35.2). Their sketches were another step toward planning a physical project. Students were aware they would not be bound to their sketches; it was simply a way to explore ideas before building (disciplinary core idea ETS1.B: Developing Possible Solutions; NGSS Lead States 2013).

Before digging through Lego pieces and starting construction, students were put into teams of two or three. To do this, the teacher surveyed the class and determined which students were

FIGURE 35.2

Engineering Planning Sheet, Student Example

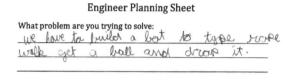

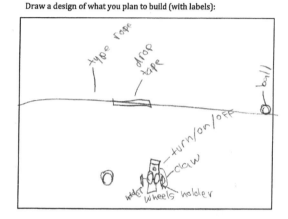

Build it—Test it—Make it better

eager to attempt a complex achievement level and which students felt more comfortable working on a less complex design. Some students wanted to reach the highest achievement level. And some wanted to try just the basics. Organizing the students into teams based on shared achievement goals helped avoid larger conflicts during the project. Later, it would be easier for the teacher to offer differentiated instruction and encourage each group to reach for a higher achievement level.

In their groups, students needed to complete one more task before the building process. Each person in the group had come up with a different sketch. The small groups discussed their sketches and built consensus around a final design and achievement level they would pursue. In many cases, individual students had similar features in their design that the group decided to keep.

For example, most students sketched a hanging vehicle and all final designs for this project ended up suspended beneath the rope. The claw that retrieved the ball had different designs. Some students proposed a pincer method. Others proposed a scoop method.

When building consensus, students made comments such as, "The aim would have to be perfect for that to work" for a pincer design or, "There's more room for error" for a scoop design.

Keep in mind that part of the planning process should include determining safety precautions to be taken during the building stage. Before beginning to build, remind students to wear eye protection. Ensure that the rope is secured at the ends before testing and that no fragile materials or equipment are in the rope's path. Advise students to work carefully around ropes and other potentially hazardous building materials.

Building

Students had nearly three hours over three class periods to construct their designs. During the building period, students were able to test their creations on the rope. Because the larger tightrope challenge had several smaller challenges to overcome, designs changed during the construction and testing process.

Nearly every group was fully engaged in the engineering process. Students created and tested one solution and continued on to another. However, one of the groups that had chosen to build a simpler design said, "We're finished," about halfway through the building time. It was a simple vehicle that could move on the rope. The teacher watched a brief demonstration of the project and agreed that they had accomplished part of the challenge, "But," he said, "think about the challenge as a whole. What can you add to this project to make it even better?" There was a moment of silence as the group members stared

at their project. Then, one of them said, "Oh, we can make something that carries a ball." The group re-entered the engineering design process.

During the three hours, students were very self-directed as they continued to create, test, and improve. There were several trial balloon comments between group members such as, "What if we tried this?" and "Maybe it would work better if we did that." However, this was also a time when setbacks and frustration occurred. In many cases, when confronted with a design problem, students demonstrated very good resiliency. Some students encountered setbacks and quickly dove back into a redesign. Some student groups needed teacher prompts such as the example above. In some groups, however, individual students had varied responses to setbacks.

One particularly interesting group had someone with a particularly strong growth mindset and another with a strong fixed mindset. After some setbacks, one group member had clearly become frustrated and was on the verge of giving up. Another student in the group began coaching him through the frustration. "Sometimes problems happen," she said. "What can we do to fix it?" After exploring some possible options, she was able to use her growth mindset and leadership skills to bring the student back into the design process.

Completing the Challenge

On the final day of the tightrope challenge, student groups demonstrated their projects for the class. Although they all didn't work exactly how they did during testing, all projects successfully completed some aspects of the challenge. Although some groups initially chose to do the simplest aspect of the challenge, none of the final projects were limited to the simplest form. Each of these groups added at least one additional challenge feature to their project. Despite difficulty and time constraints, one group managed to create a project that fulfilled all aspects of the project.

Assessing the Challenge

Throughout this engineering challenge, various formative assessments were implemented. Beginning in the early planning stages, the engineering planning sheet helped demonstrate a basic understanding of the design challenge. In addition to this basic sheet, students might also be asked to provide detailed labels and descriptions of each component.

In small groups, formative peer- and self-evaluation took place. Students considered multiple solutions and communicated and negotiated a final group design based on the merits of individual ideas.

During the construction process, student work was relatively self-directed. This allowed the teacher to act more as a facilitator by asking questions such as, "What else can you try?" or "Would another part make a difference?" By circulating the room, the teacher could determine who was fully invested in the engineering process and who might need some encouragement.

Once the challenge was complete, students completed a written self-assessment rubric (see NSTA Connection, p. 306). While some components of the projects were assessed, much of the rubric referenced the quality of students' interpersonal interactions.

On the rubric, students ranked themselves somewhere within three levels of achievement. The "starting" level was described to the students as not quite working up to your ability. The "building" level was on target for fourth-grade work. The "performing" level was above and beyond. Because students were able to choose the complexity of their design before beginning the challenge, the same assessment tool could be used for all student groups.

Alternatives to Legos

Not every school has Lego resources to do this particular challenge. To build engineering skills, it's not the materials that matter as much as participating in the engineering process. Learning about the engineering design process can take many forms. There's a good chance that you already implement lessons where this process would be a natural fit. Or, maybe you are just considering an engineering challenge and want to start small.

Perhaps students could design and build a tall structure using limited materials, such as wooden craft sticks, tape, and string. Students may already participate in a form of the egg drop challenge, and the engineering design process can be woven in seamlessly. There are several online and print resources that can offer ideas for engineering challenges (see Internet Resources, p. 306). But often, the best and most original ideas are engineered by a teacher's own imagination.

In addition to the written assessment, students discussed some of the challenges and successes of the project. For this challenge, we discussed limiting factors. Frequently we are limited in what we can do. We don't have everything we need to make things work the way we imagine them. Students were given the question, "What limited your group from making a perfect robot?"

Several hands went up. One student said, "We only have these Lego pieces, and sometimes they don't fit together the way I want them to."

The teacher did a quick survey, "Lego makes tons of pieces. How many people could have made a better project if we had every piece we wanted?" Most hands went up.

Another student said, "Time." Another quick survey. "How many people could have made a better project if they had another hour or even five hours?" Most hands went up.

Then, a student gave a very astute response. "Experience," he said. When asked to elaborate he said, "We're only in fourth grade. If we practice and keep learning, we'll be able to build better things" (*Common Core State Standards, English Language Arts,* K–5 Speaking and Listening; NGAC and CCSSO 2010). Clearly that last student has a growth mindset.

Conclusion

Just like other academic disciplines, learning to become an innovator takes time and practice.

Filling students with information doesn't teach problem-solving skills. When students are given open-ended challenges to solve, authentic learning takes place. They practice real-world skills such as collaboration, negotiation, and teamwork. In the safety of the classroom, they work through difficulties such as frustrations, conflicts, and setbacks. When facilitating engineering activities such as these, it's often easy to spot examples of Dweck's (2008) fixed and growth mindsets. Although growth and change often takes time, we can create an engineering atmosphere where every student experiences some level of success while learning that there's always room to improve their work.

Connecting to the *Next Generation Science Standards*

The materials, lessons, and activities outlined in this chapter are just one step toward reaching the performance expectations listed below. Additional supporting materials, lessons, and activities will be required.

3-5-ETS1 Engineering Design *www.nextgenscience.org/3-5ets-engineering-design*	**Connections to Classroom Activity**
Performance Expectations	
3-5-ETS1-1: Define a simple design problem reflecting a need or a want that includes specified criteria for success and constraints on materials, time, or cost	Described the design challenge in writing based on their understanding of the problem and their anticipated level of achievement
3-5-ETS1-2: Generate and compare multiple possible solutions to a problem based on how well each is likely to meet the criteria and constraints of the problem	Designed and sketched individually conceived solutions and compared and contrasted them with solutions of their peers
3-5-ETS1-3: Plan and carry out fair tests in which variables are controlled and failure points are considered to identify aspects of a model or prototype that can be improved	Tested the design and function of their robot and program to determine necessary changes to complete the design challenge
Science and Engineering Practices	
Planning and Carrying Out Investigations	Planned a design strategy, built a project based on the design, and carried out tests to determine the design effectiveness
Constructing Explanations and Designing Solutions	Based on designs, used text and sketches to explain a particular design idea that can complete the task

3-5-ETS1 Engineering Design *www.nextgenscience.org/3-5ets-engineering-design*	**Connections to Classroom Activity**
Disciplinary Core Ideas	
ETS1.A: Defining and Delimiting Engineering Problems • Possible solutions to a problem are limited by available materials and resources (constraints). The success of a designed solution is determined by considering the desired features of a solution (criteria). Different proposals for solutions can be compared on the basis of how well each one meets the specified criteria for success or how well each takes the constraints into account.	Collaborated on design solutions using limited Lego materials to achieve multiple objectives
ETS1.B: Developing Possible Solutions • At whatever stage, communicating with peers about proposed solutions is an important part of the design process, and shared ideas can lead to improved designs. • Tests are often designed to identify failure points or difficulties, which suggest the elements of the design that need to be improved.	Created sketches and designed plans to share and discuss with others Compared and contrasted design solutions as a team Built and tested various designs to evaluate effectiveness
ETS1.C: Optimizing the Design Solution • Different solutions need to be tested in order to determine which of them best solves the problem, given the criteria and the constraints.	Used results from testing trials to determine and implement design changes
3-PS2.A: Forces and Motion • Each force acts on one particular object and has both strength and direction. An object at rest typically has multiple forces acting on it, but they add to give zero net force of the object. Forces that do not sum to zero can cause changes in the object's speed or direction of motion.	Investigated ways in which various strategies impact the force, stability, and motion of objects
Crosscutting Concept	
Cause and Effect	Determined how constructed design and programming work together to influence robot functioning in the physical world

Source: NGSS Lead States 2013.

References

Bedi, J. 2004. Thomas Edison's inventive life. Lemelson Center, Smithsonian Institution. *http://invention.si.edu/thomas-edisons-inventive-life.*

Dweck, C. 2008. *Mindset.* New York: Ballantine Books.

National Governors Association Center for Best Practices and Council of Chief State School Officers (NGAC and CCSSO). 2010. *Common core state standards.* Washington, DC: NGAC and CCSSO.

NGSS Lead States. 2013. Next Generation Science Standards: For states, by states. Washington, DC: National Academies Press. *www.nextgenscience.org/next-generation-science-standards.*

Internet Resources

Children's Engineering Educators
www.childrensengineering.com

Engineering Design Process Graphic
www.theworks.org/educators-and-groups/educator-resources/engineering-design-process

LEGO Education Community
www.legoeducation.us

NASA Engineering Design Challenge
www.nasa.gov/audience/foreducators/plantgrowth/reference/index.html#.UtmL6PbnbWI

NSF Engineering Classroom Resources
www.nsf.gov/news/classroom/engineering.jsp

NSTA Science Store
www.nsta.org/store

NSTA Connection

Visit **www.nsta.org/SC1409** for step-by-step photos of how to build the Lego device, blank copies of the engineer planning sheet, and the self-assessment rubric.

Modeling Water Filtration

Model-Eliciting Activities Create Opportunities to Incorporate New Standards and Evaluate Teacher Performance

By Melissa Parks

So much is changing in education: the introduction of the *Common Core State Standards* (*CCSS*) and *Next Generation Science Standards* (*NGSS*), STEM learning opportunities, and in my state, the implementation of teacher evaluation using the Marzano Art and Science of Teaching Framework (Marzano 2007).The framework consists of four domains—classroom strategies and behavior, planning and preparation, reflecting on teaching, and collegiality and professionalism—used to define and systematically develop teaching expertise.

These changes have forced me to reevaluate my teaching. I needed to find a method that allowed me to integrate *CCSS* and *NGSS* into my fifth-grade classroom, while being mindful of my new teacher evaluation process. I have found a tool that enabled me to satisfy all those criteria—model-eliciting activities (MEAs).

MEAs are not new to those in engineering or mathematics, but they were new to me. MEAs are simulated real-world problems that integrate engineering, mathematical, and scientific thinking as students find solutions for specific scenarios.

During this process, students generate solutions by hypothesizing, testing, refining, and extending their thinking while creating models (Lesh et al. 2000). Through the hands-on, performance-based tasks, students manipulate, discuss, and defend their ideas in small groups. As MEAs are integrated into the curriculum, the learning experiences may increase the depth of students' understanding and lead them to apply that knowledge to new situations (Garfield, delMas, and Zieffler 2012). MEAs address, to varying degrees, the K–12 science and engineering practices outlined in the *NGSS* (NGSS Lead States 2013):

1. Asking Questions (for science) and Defining Problems (for engineering)

2. Developing and Using Models

3. Planning and Carrying Out Investigations

4. Analyzing and Interpreting Data

5. Using Mathematics and Computational Thinking

6. Constructing Explanations (for science) and Designing Solutions (for engineering)

7. Engaging in Argument From Evidence

8. Obtaining, Evaluating, and Communicating Information

MEAs integrate scientific, engineering, and *CCSS* as students discuss and write their findings in a manner that clearly communicates the results of their work.

MEAs can be used to introduce or review content. I elected to review content because students had just completed a unit on ecosystems and climate (5-LS2 Ecosystems: Interactions, Energy, and Dynamics and LS2.A: Interdependent Relationships in Ecosystems; NGSS Lead States 2013). To intertwine scientific, engineering, and mathematical practices, I created an MEA that would challenge students to build a cost-effective water filter to clean a (teacher-made) water sample of pollutants. Students' curiosity was piqued, and they actively engaged in the real-world problem solving that required multiple steps to take the project from brainstorm to building to testing and revising, all while discussing options with peers, which are practices outlined in the *NGSS* (Asking Questions and Defining Problems and Constructing Explanations and Defining Solutions). This collaborative process may be beneficial to English language learners as they integrate both social and academic language throughout the MEA.

A Peek Into a Classroom: Ranking and Justifying

I introduced the MEA by presenting a graduated cylinder filled with "polluted" water. I created the sample using food coloring, cooking oil, gravel, and dirt. I added a small plastic turtle floating in

the water. I placed the container on a desk in the classroom and announced, "We have a problem." I was immediately met with comments of, "Ehh, gross!" and "What is that?" I introduced the problem the students would be solving by explaining, "A friend of mine had a picnic along the river last weekend and was very disturbed by what she saw. She was so bothered, she took water samples and wrote us a letter asking for help" (Figure 36.1). The ultimate learning goal for this MEA was for students to create a water filter that would remove pollutants from the water sample: the debris, oil, and even the food coloring.

I arranged the students randomly into groups of four and distributed the MEA introductory problem letter. Students were instructed to read the letter, review the filter test results data chart at the bottom of the page, and discuss the following water filtration questions on the back of the paper (see NSTA Connection, p. 315) in their groups: What do you need to create to solve the problem? Who will use your solution? What things need to be included in your solution? What might be difficult about solving this problem for the user? Students were allotted 10 minutes of discussion time during which they had to record their collaborative thoughts. At this begining stage, conversations were focused on water clarity. Some students felt the water should be clear, whereas others thought particle amount was critical. I was surprised at the depth of conversation regarding particles and what they were, including attention to whether a particle was natural (such as a plant piece, which could be benefical) or manmade (such as a plastic bag, which would be harmful).

I brought the students back together and asked several groups for a summary of their responses. Clarity and particle amounts were the two most popular discussion topics. Only one group mentioned cost and supported the thought that cost of supplies matter because if the supplies were too expensive no one would

FIGURE 36.1

Introductory Challenge Letter

Dear Engineering Team,

My name is Emerson. I am a middle school student from Fort Lauderdale, Florida. Yesterday, I saw a turtle on the bank of the New River. The New River is a large river that runs through the Nature Center where I volunteer; it is part of the intracoastal waterway. I looked at the turtle and noticed a shiny rainbow in the water—oil! I was worried about the turtle swimming in oily water and decided to capture it in a container without touching it. Right now, we have the turtle in a fish bowl at the Nature Center.

But I have a problem. I know that the water in the fish bowl will not stay clean for long. I cannot change the water in the fish bowl because the New River water is dirty. I need to clean the water with a filter. I have built several water filters. I do not know how to rank the filters. I would like your engineering team to look at test results from my filter designs. The chart below describes the color of the filtered water. It also describes the particles that remain in the filtered water. I included how much each filter cost to build and how long the filter took to clean the cup of dirty water. Please use this data to help me develop a procedure to rank the filters so it does not take too long to clean dirty water. Finally, your team must decide how clean the water must be for the turtle to be safe.

Here is a list of the materials that I used to create my filters:

- Window screen
- Coffee filter paper
- Gravel
- Sponges
- Sand
- Cotton balls

Here are the results of the filter tests that I have done so far.

Feature	1 (Emerson)	2 (Emerson)	3 (Emerson)	4 (Emerson)	5 (Emerson)
Water color	light green/green	clear	light green	light brown/green	green/brown
Particles present	few small particles	particle free	few small particles	many small particles	small particles present
Cost	$5.50	$13.00	$6.75	$9.50	$6.75
Time to filter one bowl of water (minutes)	8	30	15	25	5

Please write a letter to me telling me how to rank all of my filter designs. In your letter, send in the work that shows me how your team ranked the filter designs. Don't forget to tell me why you think your way of ranking water filter designs will help me pick the best design.

Several of my friends have created filters for me. They promised to keep making new filters. Your procedure will help me compare all the filters.

Thank you for your help! If my turtle could talk, she would thank you too!

Emerson

build or use the filter. I then redirected the students to the data table on the introductory problem letter and reiterated the first part of the task. Students were asked to use the data chart provided on the introductory letter and rank which filters in the chart would be the best for cleaning the polluted water. Students had 20 minutes to discuss, debate, and rank the presented filters. At this point, students had been introduced to types of water pollution and ways of remedying it, but only in a preliminary manner based on textbook information. They had no classroom-based knowledge of the criteria for effective water filtration. Students were given no information on how the costs (which I randomly generated) were created. Conclusions on what would constitute an acceptable water filter were left to the students to determine based on the text and their personal prior knowledge. After students ranked the filters, groups were instructed to formulate justifications of the ranking using the data in the table on the introduction letter.

As students discussed, I circulated, but intentionally did not interject.

All but one group quickly came to consensus about which water filter they felt was best. When the allotted time expired, one group still had not ranked the filters. Within the group of four (two girls and two boys), the girls were having a heated debate on which was the most important factor—the clarity of the water or the cost of the filter—while the boys sat and watched.

Sara: "We need to have the filter that makes the clearest water."

Tonia: "Yes, but that one is too expensive."

Sara: "But we need to have clear water for the turtle to be healthy."

Tonia: "But who has all that money? Think about it. It takes $13 to clean the bowl. And that is only for 30 minutes."

Sara: "But it's only $13. It doesn't say we have to clean the bowl more than one time."

Tonia: "Of course you have to clean it more than one time. The turtle eats and is going to the bathroom. You have to clean it all the time."

Matthew: "What if you only had to clean the bowl once a day? Then that would be an OK first choice. It's expensive, but it works."

At this point, the group asked for my input, and I advised one filter cycle per day would be fine. I had not anticipated that level of student analysis!

To summarize the first portion of the MEA, I recorded each group's ranking on a class data chart. As each group presented their rankings, I challenged students to use the data in the chart to justify their rankings. This was a bit difficult for some students that struggled with the idea of needing to support their rankings: "We thought the clearest water is the best for the turtle so filter two was our first choice." Other students excelled at using the data in their justifications, "Because turtles don't live in crystal clear water, we went with how fast the water could be cleaned, so we picked filter five. And turtles live in water that has stuff in it all the time and they eat some of that stuff, and we wanted to be able to clean the water quickly, but not make it like a swimming pool." After reviewing each group's rankings, I told the class we would resume this activity the next day.

Building, Testing, and Revising

After reviewing the previous day's rankings, students were instructed to create a letter to Emerson explaining their rankings of the filters on the introductory problem letter. Then, students were challenged to create a filter using the materials listed: cotton, sponges, coffee filter paper, sand, gravel, and window screen (see NSTA Connection, p. 315). Throughout the building process, students wore splash goggles. I also watched for slip hazards from spills and made sure that students washed their hands after using the materials.

Students had 45 minutes to brainstorm, build, test, and revise a water filter using the criteria each group determined previously to be the most important (Asking Questions and Defining Problems; ETS1.A: Defining and Delimiting Engineering Problems; and ETS1.B: Developing Possible Solutions; NGSS Lead States 2013).

For example, a group that determined water should be particle-free would strive to build a filter that would reduce or eliminate the particles in the water. A group that determined water color to be the most important criteria would build a filter that would reduce the food coloring in the "polluted" water. I did not have students diagram their filter, but upon reflection I see that would have been a valuable asset to enhancing the design cycle. Furthermore, the labeling and diagramming of the student-created filter could have presented another opportunity to link oral and written expression of learning.

At the end of the lesson, each group presented the water filter they built to the class. Filters were evaluated based on the criteria in the data chart: water color, amount of particles, and length of time to filter the water. Challenging students to express themselves orally and on paper are parts of the *CCSS* for speaking, listening, and writing (CCSS.ELA-LITERACY.SL5.1a-d, 5.4; CCSS.

ELA-LITERACY.W.5.1–5.2; NGAC and CCSSO 2010). During the group presentations, I created a second class data chart on which students' data were entered. According to feedback from the students, this part of the MEA could easily have stretched to double the 45 minutes allotted because of students' enthusiasm and involvement. One student shared, "This is fun. I wish we had more time to work on our filter. It took a lot longer than we thought to pour the water through the filter. If we had more time, I think we could have had clearer water."

Reviewing and Synthesizing

Students were surprised their "clean" water looked different after 24 hours. Some had particles of sand settle on the bottom of the container. Others had a layer of oil on top of their sample. This was a perfect, although not planned, review of mixtures and solutions—a by-product of this MEA. On day three, students were presented with a twist in the MEA (see Figure 36.2, p. 312). I inserted two additional filter types into the MEA problem letter. Students were now asked to synthesize their personal filter building from the previous day with this new information and rerank the filters (Constructing Explanations and Designing Solutions).

Students had 45 minutes to rerank the new list of filters and prepare an oral and written justification of the process. Again, I circulated but did not provide guidance. My job was to facilitate discussions. If students were stuck on a particular point, I would try to move the discussion along by asking probing questions: "Did you talk about the comparison of speed and cost?" and "Did you think about what could be more important, the clarity of the water or the amount of particles in the water?"

Throughout this multiday activity, I formatively assessed students using a rubric with a scale of

FIGURE 36.2

Challenge Letter With Two New Filters Added

Dear Engineering Team,

You must have worked very hard! I received your letter telling me how to rank my filter designs to clean the water in the rescued turtle's bowl. Your procedure was very helpful. That is good because I need more help from your team!

My friends came up with two more filter designs to help me clean the dirty water in the bowl. I have included my filter design results in the data table with the new results. The new results are under the numbers "(Friend) 6" and "(Friend) 7" in the table.

Feature	1 (Emerson)	2 (Emerson)	3 (Emerson)	4 (Emerson)	5 (Emerson)	6 (Friend)	7 (Friend)
Water color	light green/ green	clear	light green	light brown/ green	green/ brown	light green	green
Particles present	few small particles	particle free	few small particles	many small particles	small particles present	particle free	small particles
Cost	$5.50	$13.00	$6.75	$9.50	$6.75	$9.50	$5.00
Time to filter one bowl of water (minutes)	8	30	15	25	5	10	10

Write another letter to me explaining how to rank all of the filters. In your letter, send me the work that shows how your team chose to rank them. Your team might need to change the old procedure if it does not work for the new filter designs. My friends are still making more filters. I will use your team's procedure to help me compare all of the filters, both old and new.

Thank you for all your help.

Emerson

0–3 (see NSTA Connection, p. 315). The rubric was dual purpose. First, I informally assessed students throughout the MEA and recorded my assessment on the rubric. Then, at the conclusion of the MEA, students received the rubric I used to assess them and were asked to record a self-reflection of their experience on the same rubric. I found this self-assessment to be very informative. The data allowed me to monitor my awareness of the learning in the classroom. Was there agreement between my assessment and the students' self-reflection? If a particular issue was discussed, did I observe the same issue? If not, why not?

The biggest challenge I have found for implementation of MEAs is the time commitment to implement it properly. To be effective, students need time to process information, apply, and revise their thinking based on results and group discussions. This takes multiple days and lots of patience. I found it very difficult not to insert my point of view in an effort to guide the students. The struggle is part of the MEA process, which challenges students to broaden their thinking. Additionally, students have not had formal training on how to argue or persuade, and that led some groups to get stuck in a repetitive cycle of not wanting to concede a point and being unable to articulate thoughts with enough detail or persuasion to convince others. This led to time quickly running down while students were stuck in these patterns. As a result, students quickly agreed on one solution to satisfy my time restraints and then were unable to justify why they chose that particular option either in creation or ranking. One group expressed, "I hated the fact we were rushed for time. Because we hurried, our letter was only OK. Next time, we'd like to have more time to talk about our work."

This class has encountered both joy and frustration while completing the MEA. The class enjoyed the collaborative and sometimes competitive aspects of working in groups. They also felt the frustration of having a finite amount of time to complete the task each day. I have only begun to use MEAs as part of my science curriculum and can see the potential for greatness in their application. I find the MEAs to be a particularly effective way to integrate multiple standards—both the *CCSS* and the *NGSS* (English language arts, mathematics, and science) in meaningful ways.

The discourse between students (informative, persuasive, and argumentative) meets the *CCSS* literacy standards of speaking and listening while the letter writing meets the *CCSS* writing standards. One student reflected, "I loved the drama of being an engineer. Our group had some challenges about which filter would be the best, but we eventually all worked together to come up with the best design to help the turtle." I like the idea that using MEAs in my class affords my students the opportunity to explore a new way of problem solving that requires them to plan, build, test, and redesign. The integration of MEAs into an already busy school day is not easy, but is well worth the effort based on the levels of engagement, learning, and enthusiasm I've seen in the classroom.

Connecting to the *Next Generation Science Standards*

The materials, lessons, and activities outlined in this chapter are just one step toward reaching the performance expectations listed below. Additional supporting materials, lessons, and activities will be required.

3-5-ETS1 Engineering Design www.nextgenscience.org/3-5ets1-engineering-design **3-PS2 Motion and Stability: Forces and Interactions** www.nextgenscience.org/3ps2-motion-stability-forces-interactions	Connections to Classroom Activity
Performance Expectations	
3-5 ETS1-1: Define a simple design problem reflecting a need or a want that includes specified criteria for success and constraints on materials, time, or cost	Collaborated, brainstormed, and explained which type of water filter would fulfill the need based on materials, cost, and time
3-5-ETS1-2: Generate and compare multiple possible solutions to a problem based on how well each is likely to meet the criteria and constraints of the problem	Built and tested water filters and compared results using provided constraints including material, cost, and time to filter
Science and Engineering Practices	
Asking Questions and Defining Problems	Posed, responded to, and deliberated questions about the best options for creating a water filter
Planning and Carrying Out Investigations	Built the water filter, tested and revised the design based on clarity of water with each test
Constructing Explanations and Designing Solutions	Explained and justified filter designs to both teammates and classmates (whole group) using data collected during tests
Disciplinary Core Ideas	
ETS1.A: Defining and Delimiting Engineering Problems • Possible solutions to a problem are limited by available materials and resources (constraints).	Aligned design and justification of final design based on provided constraints (materials, cost, time)
ETS1.B: Developing Possible Solutions • At whatever stage, communicating with peers about proposed solutions is an important part of the design process, and shared ideas can lead to improved designs.	Participated throughout the planning, testing, revising, and presenting stages in discourse including discussions, explanations, arguments, and rebuttal ideas based on evidence in the activity
PS.2.A: Forces and Motion • Each force acts on one particular object and has both strength and a direction.	Designed filters that allow some liquids to pass through
Crosscutting Concept	
Influence of Science, Engineering, and Technology on Society and the Natural World	Created a water filter to improve the quality of water in a teacher provided, polluted sample

Source: NGSS Lead States 2013.

References

Garfield, J., R. delMas, and A. Zieffler. 2012. Inventing and testing models: Using model electing activities. The Science Education Resource Center at Carleton College. *http://serc.carleton.edu/sp/cause/mea/index.html*.

Lesh, R., M. Hoover, B. Hole, A. Kelly, and T. Post. 2000. Principles for developing thought-revealing activities for students and teachers. In *Research design in mathematics and science education*, ed. A. Kelly and R. Lesh, 591–646. Mahwah, NJ: Lawrence Erlbaum Associates.

Marzano, R. 2007. *The art and science of teaching: A comprehensive framework for effective instruction*. Alexandria, VA: ASCD.

National Governors Association Center for Best Practices and Council of Chief State School Officers (NGAC and CCSSO). 2010. *Common core state standards*. Washington, DC: NGAC and CCSSO.

NGSS Lead States. 2013. *Next Generation Science Standards: For states, by states*. Washington, DC: National Academies Press. *www.nextgenscience.org/next-generation-science-standards*.

NSTA Connection

Find the challenge letters, water filtration questions worksheet, filter ranking letter template, and rubric at *www.nsta.org/SC1410*.

Index

Page numbers in **boldface** type refer to figures or tables.

Index